MW01519064

The Business of Conquest

The
Business of Conquest

EMPIRE, LOVE, AND LAW IN THE ATLANTIC WORLD

NICOLE D. LEGNANI

University of Notre Dame Press
Notre Dame, Indiana

Library of Congress Control Number: 2020946986

ISBN: 978-0-268-10896-0 (Hardback)
ISBN: 978-0-268-10899-1 (WebPDF)
ISBN: 978-0-268-10898-4 (Epub)

Para mi hija, Francesca Delia

CONTENTS

ILLUSTRATIONS

ACKNOWLEDGMENTS

Perhaps it is fitting that a study that reckons with the moral and material debts incurred by various agents in the Iberian conquests should participate in the long-established trope of the author expressing her gratitude for the support received from people and institutions, without whom and without which this book would have been impossible to complete. So I begin with my thanks for the genre itself, which designates this space at the beginning to itemize both my outstanding debts and my sincere assurances of my intention to repay them, alongside the deep-seated conviction that a commensurate settlement remains an impossible but not for this reason, less indispensable task to undertake in the lines below.

I am deeply indebted to Eli Bortz, my editor at the University of Notre Dame Press, and his editorial team for their unflagging support for and meticulous work on this project. Sheila Berg, you are a rock star.

It has been a privilege and a pleasure to benefit from two especially generous mentors and rigorous readers over the years, José Rabasa and Mary Malcolm Gaylord. Their patience, sense of humor, and encouragement never fail to surprise and guide me.

To José Rabasa, who first directed me with marginal notes and questions and in phone conversations and meetings over coffee ever since I was a graduate student, my deepest gratitude for continuing to read me and for providing me with invaluable advice such as "No te comas el coco." He continually challenges me to embrace the questions that arise in writing, questions that must be raised precisely because they have no easy answers.

This book would not have been possible without the unparalleled support and nurturing of Mary Malcolm Gaylord, who remembers where I sat in all the seminars that she taught and has never failed to support my professional development, who cooks for and hosts dinners at her home in Concord, and who asks variations on the question "So what?," as needed, in copious marginalia. She has been my teacher, mentor, and friend since I was a first-year student at Harvard College. Her attention to detail, argument, and structure is without parallel. My thanks for her patience and support and for believing in the salience of this project from the outset.

I am extremely grateful to Juan Vitulli for his extensive commentary, questions, and suggested revisions of the manuscript. I am similarly grateful to the anonymous reviewer for extremely positive feedback and suggestions. I cannot begin to express my thanks to Jorge Téllez Vargas for reading and commenting on earlier versions of the introductory and third chapters and to Isis Sadek for her insightful comments and recommendations on later versions. My thanks also to Andy Alfonso, Juan Diego Pérez, and Robert A. Myak, whose sharp eyes and attention to detail were integral to the editing and proofreading processes.

I would also like to extend my deepest gratitude to José Antonio Mazzotti for recognizing my vocation when I was an undergraduate and for encouraging and supporting my first foray into colonial Latin American studies.

I must also thank my colleagues with whom I have the great pleasure of working in the Department of Spanish and Portuguese at Princeton University: Marina Brownlee, Alberto Bruzos Moro, Nicola Cooney, Arcadio Díaz Quiñones, Rubén Gallo, Javier Guerrero, Germán Labrador Méndez, Christina Lee, Angel Loureiro, Pedro Meira Monteiro, Gabriela Nouzeilles, Rachel Price, and Ron Surtz. Though Bruno Carvalho has left us, I would certainly be remiss if I failed to thank him as well. Fernando Acosta, our curator, and Gabriel Swift, at the Rare Books and Special Collections in Firestone Library, fielded all my questions and last-minute requests. I would also like to thank Mitra Abbaspour at the Princeton Art Museum for reaching out to me in the spring of my first year at Princeton when I was teaching my first graduate seminar. Without her inquiry and initiative, I would not have met the Postcommodity collective, Raven Chacón, Cristóbal

Martínez, and Kade L. Twist, whose praxis continues to be an inspiration. To Vera Candiani, the ardent inquisitor of my use of the term "venture capital," my thanks for the pushback and the camaraderie. I am also grateful to Sarah Rivett for opening a space for Indigenous studies to flourish at Princeton, the settled and unceded territory of the Lenni-Lenape.

My thanks to Danelle Gutarra Cordero for organizing the Postcolonial Humanities Working Group through the Humanities Council at Princeton, where I benefited greatly from participants' questions and comments on Las Casas and his Madeira rabbits. I am also grateful to the Tepoztlán collective for affording me the opportunity to present two papers during the 2014 and 2017 summer meetings. The theory read, commentary given, and performances presented by participants informed the sections on José de Acosta in the introduction and chapter 4 and the Las Casas section in chapter 3. I also wish to thank Tulia G. Falletti and Cathy Bartch for inviting me to give the keynote address at the third annual Penn in Latin America and the Caribbean conference in October 2017. The longer format allowed me to trace the relationships between Las Casas and Columbus as developed in chapter 3.

I would also like to acknowledge the editors of *Latin American Culture and the Limits of the Human*, who permitted me to reproduce sections from my chapter, "Invasive Specie: Rabbits, Conquistadors, and Capital in the *Historia de las Indias* (1527–1561) by Bartolomé de Las Casas (1484–1566)," in chapter 3. These selected excerpts are reprinted with the permission of the University Press of Florida from my chapter in *Latin American Culture and the Limits of the Human*, edited by Lucy Bollington and Paul Merchant (Gainesville: University Press of Florida, 2020).

Over the years, many of the students who sat around a seminar table with me have proved influential in my research and writing process. When I served the Romance Languages and Literatures Department as a College Fellow at Harvard University, I was especially grateful for the insights provided by James Almeida, Henry Brooks, José de León González, and Wilnomy Pérez Pérez. As an assistant professor in the Department of Spanish and Portuguese, I have been similarly blessed to sit at a roundtable with Andy Alfonso, Luisa Barraza Caballero, Vero Carchedi, Berta Del Río Alcalá, Yangyou Fang, Jannia

Gómez González, Ryan Goodman, Alejandro Martínez Rodríguez, Sean McFadden, William Mullaney, Juan Diego Pérez, Paula Pérez Rodríguez, Paulina Pineda, Sowmya Ramanathan, Margarita Rosa, and Peter Schmidt.

Over the years, I have benefited from conversation with and camaraderie of friends, readers, and co-presenters who have informed my research and writing in creative and productive ways: Arantxa Araujo, Santa Arias, B. Chrissy Arce, Antonio Arraiza, Orlando Bentancor, Josiah Blackmore, Monique Blom, Lotte Buiting, Luis Cárcamo Huechante, Rodolfo Cerrón Palomino, Enrique Cortés, Gregory Cushman, Jessica Delgado, Ivonne Del Valle, Susana Draper, Caroline Egan, Luis Girón Negrón, Goretti González, Karen Graubart, Evelina Guzauskyte, V. Judson Harward, Michael Horswell, Rosario Hubert, Nick Jones, David Kasanjian, Stephanie Kirk, Salomon Lerner Febres, Obed Lira, Melissa Machit, Yolanda Martínez-San Miguel, Kelly McDonough, Michelle McKinley, María Rosa Menocal (*q.e.p.d*), Leah Middlebrook, Giovanna Montenegro, Anna More, Cristina Moreiras Menor, Chris Morin, Abdul-Karim Mustapha, Afsaneh Najmabadi, Dan Nemser, Sophia B. Núñez, Simone Pinet, Rachel O'Toole, María Josefina Saldaña Portillo, David Sartorius, Sarah Winifred Searle, Mariano Siskind, Daniel Strum, Analisa Taylor, Zeb Tortorici, Carlos Varón González, Miguel Valerio, Manuela Valle-Castro, Sonia Velázquez, Luciana Villas Bôas, Pamela Voekel, Lisa Voigt, Dillon Vrana, Emily Westfal (RIP), and Gareth Williams. I would also like to extend my heartfelt thanks to my fellow ASADISTAS—Miguel Martínez, Aude Plagnard, Víctor Sierra Matute, Lorena Uribe, Felipe Valencia, and Juan Vitulli—for continuing to provide inspiration on a quasi-daily basis within a close-knit community founded on a shared appreciation for early modern poetry and the politics of its reception.

To Josefina Legnani, my mother, who has been battling stage 4 breast cancer since fall 2016, thank you for fighting the good fight. To Augusto Legnani, my father, thank you for reading to me every night when I was a child and for suffering my insufferable polemics about the merits and demerits of the Roman Empire.

To Jessica Fowler, who has shown me true friendship, as only a *Dictablanda* could, my deep gratitude for the morning coffees, evening drinks, and exploratory mission of Gijón, with its many coves and feminist, anarchist bookstores.

A Rafael SM Paniagua mil gracias por enseñarme el valor de un clavel y por compartir tantas e innumerables cosas y experiencias conmigo, incluyendo un salón de clases, una corrida y varios paseos a las orillas del Tajo.

Mis agradecimientos a Brunella Tedesco, primita, no hay mejor persona con quien conocer Madrid la madrugada de un domingo veraniego en plena huelga de taxis.

I am grateful to Angélica Serna Jeri, whose deep knowledge of the *huacas* of Peru never ceases to amaze me, for always being there for me and for family, no matter the distance. Mi amiga del alma con quien siempre puedo compartir algunos versos de Vallejo y citas de Arguedas, gracias por tu escucha siempre tan generosa.

To Christopher M. Morse and Thomas Gareau, whose hospitality and friendship I cherish, thank you for choosing to be family; for managing to be here in our darkest hours, even as an ocean and a continent have seemed to raise up an insurmountable distance between us.

And in the home stretch, my thanks to Jay P. Outhier, for all the love, laughter, and many books received as a present or on loan and also for sharing the "imagination of wings" and other things with me.

Finally, my deepest gratitude to Fernando Gamio, Lyanna Gamio, and Kerry Ann Sass, for enveloping me with all the love, support, and many laughs at their disposal; for providing a home away from home; and for insisting that family weekends are sacred.

And last but certainly not least, to my sassy, intelligent, generous daughter, Francesca Delia Gamio-Legnani, who never fails to inspire me with her quick wit and astute observations, my thanks to you for being who you are and for allowing me to love you wholeheartedly and unconditionally.

Introduction

Setting Sail with Felipe Guamán Poma de Ayala (1550–1616)

> A figure is (already) a little fiction, in the double sense that it
> usually takes but a few words, or even one, and its fictional
> character is mitigated by the smallness of its vehicle, and, often,
> by the frequency of its use, which prevents the perception of the
> audacity of its semantic pattern: only use and convention make
> us accept as commonplace a metaphor such as "hold a torch for
> someone," a metonymy such as "drink a glass," or hyperbole such
> as "die of laughter." The figure is an embryo, or, if one prefers, a
> sketch of fiction.
> —Gérard Genette, *Métalepse: De la figure à la fiction*

> The conquistadors undertook the Conquest at their own risk; in
> a way, it was a private undertaking. But it was also an imperial
> enterprise.
> —Octavio Paz, *Sor Juana, or, The Traps of the Faith*

At first sight, the depiction of six conquistadors on the same boat can be
disconcerting (fig. 1). The title of this image gracing the chapter on conquest in *El primer nueva corónica y buen gobierno* (1615/16) by Felipe
Guamán Poma de Ayala (ca. 1535–after 1616), a Yarivilca of Huamanga in Peru, reads, "Conquista. Embarcáronse a las Indias" (Conquest. They Set Sail for the Indies).[1] By representing the enterprises
of Christopher Columbus (?–1506), Juan Díaz de Solís (1470–1516),
Diego de Almagro (ca. 1475–1538), Francisco Pizarro (1476?–1541),

1

Figure 1. Felipe Guamán Poma de Ayala, "Conquista. Embarcáronse a las Indias." From left to right, Columbus, Juan Díaz de Solís, Almagro, Pizarro, Vasco
Núñez de Balboa, Martín Fernández de Enciso. Courtesy of the Royal Danish
Library, GKS 2232 kvart: Guamán Poma, *Nueva corónica y buen gobierno* (ca.
1615), page [373 [375]].

Vasco Núñez de Balboa (1475–1519), and Martín Fernández de Enciso (1470–1528) on one boat and one visual plane, and, by extension, beyond the frame, to the lands, seas, and peoples "discovered" by these voyages of conquest during the first half of the sixteenth century, Guamán Poma allegorically abbreviates the many ventures referred to as the Spanish Conquest.[2] In doing so, Guamán Poma also asks his audience to reflect on the relationships embodied by these men on the page and their connections to the ventures they perpetrated and represented. While most readers may apprehend that these figures serve as a visual shorthand for the Spanish Conquest, the question of *how* and *why* these six men do so encourages an inquiry into the roles they played in the conquest and whether they were related to one another in their respective lives or whether their figural connections are purely of a symbolic order, made by the author. In other words, why are they all on the same boat? Why these men specifically? What, if anything, connects each figure to the others and thus to the larger imperial enterprise beyond the frame?

One figure is dispensable; the rest are not. In an almost identical drawing, which appears earlier in the chronicle (fig. 2), Guamán Poma omits the figure of Martín Fernández de Enciso. This occurs in a chapter that narrates the papal reigns chronologically. This chapter, "Flota pontifical de Colón en la mar a las yndias del Piru" (Pontifical Fleet of Columbus to the Indies of Peru), offers an allegory that is almost identical to the depiction of "Conquista. Embarcáronse a las Indias," which tells of the conquest of Peru during the sixteenth century. In retrospect, when the two visual allegories employing the figure "men on the same boat" are juxtaposed, the omission of Fernández de Enciso in the "Pontifical Fleet" allegory feels right; the five figures on the boat could be the five fingers on a hand; the sixth figure, Fernández de Enciso, tacked onto the poop deck of the boat on folio 373, feels like a supplement, an unwieldy appendage.[3] In Guamán Poma's visual allegory, Fernández de Enciso is thus expendable in the "Pontifical Fleet" but indispensable to "Conquest," an important distinction, lest Guamán Poma's intended royal interlocutor, Felipe III (r. 1598–1621), believe otherwise. There is something odd in the even-numbered slate of figures allegorizing the conquest in figure 1. Enterprises undertaken with the authority of the church, Guamán Poma seems to say, were not the same as those performing the *conquista*. They seem identical but not quite.

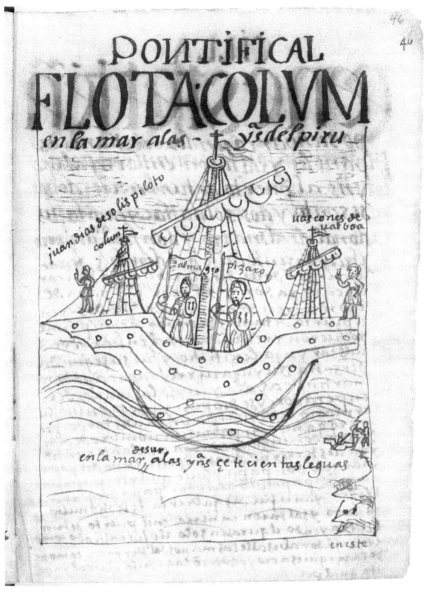

Figure 2. Felipe Guamán Poma de Ayala, "Pontifical Flota Colum en la mar a las Yndias del Pirú." From left to right: Juan Díaz de Solís, Columbus, Almagro, Pizarro, Vasco Núñez de Balboa. Courtesy of the Royal Danish Library, GKS 2232 kvart: Guamán Poma, *Nueva corónica y buen gobierno* (ca. 1615), page [46 [46]].

In his treatment of Fernández de Enciso, Guamán Poma emphasizes this conquistador's role in providing an apologetics for empire, as a tailwind to the corporate enterprise of conquest. In 1519 he wrote the *Suma de geografía*, a lesser-known work now but one translated and cited by Richard Hakluyt (1553–1616) and Francis Bacon (1561–1626) for the wealth of cultural geography it imparted. Fernández de Enciso also played a prominent role as royal geographer in the Casa de Contratación—the House of Contracts in Seville—and was one of the legal scholars thought to be behind the writing of the *Requerimiento*, the script used to perform the conquest after 1513 in the New World. The erasure of Fernández de Enciso from the "Pontifical Fleet of Columbus" allows Guamán Poma to drive an emblematic wedge in the partnership between the crown and the church in the conquest. The nature of that partnership and the resistance to the legal fictions generated by this joint venture are the subject of this book.

The manuscript of the *Nueva corónica y buen gobierno*, held at the Royal Library of Denmark in Copenhagen, narrates the times of the pre-Inca, the Inca, and the Spanish Conquest and colony and prescribes remedies for the ills and injustices of the Spanish Empire. Guamán Poma makes a logical leap from the first half of the title, the "new chronicle" of the past, to the second, with its prescriptions for good government, in that the future's potential to remedy the errors of the past depends on his own, *novel* presentation of past events. It is but one instance of the figure of *metalepsis*, broadly understood, since Aristotle defined it in his *Poetics* in the fourth century BC, as the employment of one word for another, a transference of meaning that comprised the use of figurative language, especially synonymy, metonymy, and metaphor. Taking my cue from Guamán Poma, *The Business of Conquest* explores the movement from figure to fiction in discourses of capital and violence and argues that they cannot be reduced to any one figure; instead, conquest casts a wide net, yet the fact of its artifice does not make its effects on the lives and livelihoods of the peoples known as Indians any less visceral. Logical fallacy notwithstanding, metalepsis produces powerful fictions.

"Conquista. Embarcáronse a las Indias" occurs in the section of the *Nueva corónica* that acts as a hinge between the before and the after of contact with the Spanish conquistadors and thus, indirectly, with the sovereigns of Spain. Guamán Poma goes to great lengths to separate the times of the (first) contact with Christ's apostle, Saint Bartholomew, from that (later) contact with Spanish Christians. The Spanish Conquest, Guamán Poma contends in his letter and manual to the Spanish sovereign, was an *empresa*, in the way that his own drawing was an empresa—in Spanish, both an emblem and an enterprise—and, in Guamán Poma's visual allegory for conquest, an empresa of an empresa, that is, a meta-empresa, or an emblem for an emblematic enterprise. For Guamán Poma, and other figures of the counterconquest such as the friars Bartolomé de Las Casas (1484–1566), Domingo de Santo Tomás (1499–1562), Bartolomé de Vega, and Francisco de la Cruz (1529–1578), by the 1560s the conquest had become synonymous with a business venture. An act of apostasy, it had confused Christian *caritas*, love and charity for neighbor (and, thus, God), with *cupiditas*, greed and lust. Cupiditas, suggests Guamán Poma, as did Las Casas and other brethren in Christ before him, could not be a figure for caritas or vice versa, yet such was the practice of Christianity in the Indies. However, Guamán Poma's discursive push for rhetorical consistency would land him in his own contradictions. While telling the story of the Spanish Conquest, Guamán Poma (2001) would also famously declare that there had been no conquest, given that there had not been any armed resistance (fols. 549–50 [563–64]). Thus Guamán Poma offers a contradictory narrative: the story of a conquest that was not, in fact/ish, a conquest.[4]

How can we tell the story of conquest and at the same time assert there was no conquest? Is the Spanish Conquest a nonevent, to repurpose the event as used by Badiou (2007) to name a fundamental rupture that reveals a "truth," which can be both named and unnamed?[5] If so, how shall it be named? By whom? When is this event? What is it called? Guamán Poma will both use the word *conquista* to name the event and deny its existence. "No hubo conquista" (There was no conquest) (Guamán Poma 2001, fols. 549–50 [563–64]), he will assert just as strongly in his narrative, written in alphabetic script, as he will write and demonstrate its happenstance visually in "Conquista. Embarcáronse a las Indias." Do the assertion and the negation exist in contradiction in the *Nueva corónica y buen gobierno*? I contend that the logical and narrative impasse created by the assertion and its denial serves to contest the most basic but fraught questions posed by scholars: What is a conquest? How do we both tell and contest the story of the conquest? Moreover, when and where does the story begin and end? Though Guamán Poma's "Embarcáronse a las Indias" narrates conquest in the preterit tense, a closed action, closed off in the past from our present, the "buen gobierno" section implies that the conquest, which was no conquest, was nonetheless an ongoing activity, constitutive and constituted, with no closure either for the Indigenous peoples of the Andes or for their invaders.[6]

As a capitalist empresa (enterprise but also sigil), the conquest performed a multiplying effect, in material and figurative terms. As allegorized by Guamán Poma, the series of events, known as the conquista, depended on a series of enterprises in profitable violence, the conquistas, and their *empresarios*, the conquistadors. The success and resulting excess of one conquista funded more conquistas. Yet, rather than metonymy, the preferred trope for narrating the relation among many conquests is the metaphor of conquest. On the basis of similitude and transference, through the commonplace of the *translatio, imperii*, and *studii, las dos Españas, los dos Santiagos, Matamoros y Mataindios*, or the ubiquity of SPQR inscriptions and its variations, empire reinforces its hegemony through expansion by metaphor.

By underscoring the empire's reliance on *contiguity*, which, like *contingency*, derives from the Latin verb *contingere*, "to touch," I call attention to the dependency of the Spanish empresa on bare life. In referencing Giorgio Agamben's (1998) concept of the *Homo Sacer*, I assert that with the category *indio* (Indian), imperial enterprise circumscribed toiling bodies to whom coveted lands belonged, bodies that housed the souls of potential neophytes in this hemisphere. Through the figure of Indian consent, this circumscription also worked as circumlocution in that it traced a ring of indeterminacy around those who were included in the political through their exclusion and sought to hush any recognition of the empire's limits, a precariousness exposed—paradoxically—by the empire's very dependence on Indian bodies.

Today, desired growth in enterprise is termed "scalability" and understood as the ability to replicate by projecting similitude (or expansion by metaphor). However, scalability *functions* by capital multiplying but then disassociating from the first enterprise in a contiguous form. Consequently, *empresa as enterprise* functions as a metonym, but *empresa as sigil* signifies through metaphor. To rephrase the title of

Raymond Carver's beloved book of short stories, what we talk about when we talk about the enterprise of conquest in the Indies involves the tropes, often the same ones, used both to envision the experience of the conquest and to make truth statements about what the conquest entailed.

Guamán Poma's representation of conquest by a particular set of six men recalls the contingency of these enterprises on the familiar connections among conquistadors and the dependency of each enterprise on the profitability of earlier enterprises. Recalling Quintilian's classic definition of allegory, the force of this trope (and its ironic implications) resides precisely in its literal readings. Beneath the waves, an annotation below the ship in both drawings underscores the abbreviating and totalizing vision of conquest and elucidates for Guamán Poma's readers, especially his royal interlocutor, Felipe III of Spain, that we are indeed engaging with an allegory for the transatlantic and transcontinental crossings made by these six men. Why these six men? While they all represent various stages of the Spanish Conquest in the late fifteenth and early sixteenth centuries, so too are they united by partnerships in the area of the Darién (shared by present-day Colombia and Panama), part of the geographic area first named Tierra Firme (the Continent) by Christopher Columbus on his third voyage in 1498. This area to the west of Venezuela would later be baptized Castilla de Oro (Golden Castile) according to Fernando II of Aragon's instructions to Governor Pedro Arias Dávila (also known as Pedrarias Dávila) and his cohort, who would eventually include Diego de Almagro and Francisco Pizarro, as well as Vasco Núñez de Balboa and Martín Fernández de Enciso, all of whom were involved in earlier foundational moments in the Isthmus of Panama. In fact, on the poop deck of Guamán Poma's ship of conquest, Fernández de Enciso stands behind the stowaway he discovered and pardoned on the ship under his command to Nueva Andalucía, then governed by Diego de Nicuesa (ca. 1478–1511). That stowaway, Balboa, was intent on leaving behind his debts on the island of Hispaniola by secretly joining the expedition sent to aid Alonso de Ojeda (ca. 1468–1515), who at the time shared the governance of Tierra Firme—divided between east and west—with Nicuesa. Balboa would go on to "discover" the Southern Sea (Pacific Ocean).

Balboa's connections to Pizarro and Almagro similarly originate in the Darién's hub of enterprising conquistadors via Ojeda and Fernández de Enciso. On the isthmus, Pizarro was a common soldier in Ojeda's employ, defending the new settlement of San Sebastián de Urabá, future site of Cartagena de Indias, though by the mid-1520s he, his brothers, and Almagro (who arrived in the Darién with Dávila) would have accumulated enough wealth and expertise to acquire the financial backing of Fernando de Luque (?–1533), a priest also centered in Panama, and Gaspar de Espinosa (1483–1537). Almagro, Pizarro, and Luque formalized their partnership by creating the Compañía de Levante for the conquest of Peru in 1526.[7]

In "Conquista. Embarcáronse a las Indias," these six men are all on the same boat, as it were, because of their shared ties of capital and experience in Tierra Firme, the launching site for the conquest of Peru and the earlier discovery of the River Plate. Taken together, the voyages of these six men and their business partners delineate the South American continent as a whole, rendering the continent an island. A continent circumnavigated by the collective enterprise of conquista is referenced by the measurement "la mar del Sur setecientas leguas al Río de la Plata" (seven hundred leagues from the Southern Sea to the River Plate) given in both drawings (see figs. 1, 2). Thus in Guamán Poma's visual narrative of conquest, he displays a playfulness when alluding to the connections between business partners and the lands they conquered among them. Inasmuch as the six men would seem to signify a linear progression in time, the time of conquest, a teleology of the inevitable—we all know how the story ends after all—they also define a continent by their seafaring perambulations along the northern coast of South America, south along the coast of Brazil to the River Plate, and across the Isthmus of Panama to the Southern Sea, to Peru, and beyond. Indeed, the Portuguese sailor Ferdinand Magellan (1480–1521), not pictured, famous for attempting to circumnavigate the globe with Juan Sebastián Elcano (1476–1526), chose to follow the southern course set by Solís in his earlier, ill-fated voyage that was ended in the River Plate by Tupi-Guaraní who practiced anthropophagy. Between Columbus and Solís, we might also expect the figure of Vicente Yáñez Pinzón (ca. 1462–ca. 1514), who was Columbus's business partner in his first voyage and captain of the *Niña* and later Solís's partner in the trip south along Brazil's eastern coast. To Pinzón, Peter Martyr d'Anghiera (1457–1526) first attributed the discovery of Brazil in a letter to Cardinal Luigi D'Aragona (1474–1519).[8] Via omission and reiteration, Guamán Poma's ship of conquest figures as metaphor and metonymy; the two are confused as time, contiguity, and dependency collapse in the emblematic enterprise of representing profitable violence.

The ship, much like Guamán Poma's ship shown in figures 1 and 2, has been traditionally interpreted, depending on the context, as either the ship of state (as in the opening lines of the *Aeneid*) or the church by whose good graces the ship of souls crosses the dangerous, profane seas to salvation. (Thus, *nave*, Latin for "ship," is the name given to the main body of a church.) The ship's sails have also been metonymic for desire in love lyric. I argue that in the sixteenth century the scale of the conquest of America permitted and was permitted by the synonymous pairings of "ship" so that its meanings were no longer multivalent but metaleptic. Indeed, "ship" was no longer a figure for the church *or* the state *or* greed and lust but could signify all these at once, because their significations had become synonymous, especially in the Laws of the Indies (see ch. 2) but also in theological treatises, such as the influential *De procuranda indorum salute apud barbaros* (1589) by José de Acosta, explored at length in chapter 4 in contrast to the life's work of Las Casas.[9]

Figures 1 and 2 also suggest a space for conquest that is coterminous with seafaring, beyond the demarcations of land that are at the root of the law. As scholars of early modern empires such as Lauren Benton in *A Search for Sovereignty* (2009) and Josiah Blackmore in *Moorings* (2002) have done before me with reference to Carl Schmitt's *Nomos of the Earth* (1985), I explore the paradox of imperial expansions and the construction of universal law through lawlessness on the open sea. What Henry Kamen (2004, 54) has described as the Spanish "business of empire," involving "the imposition of foreign, international capital and capitalists on the government" beginning with the ascension of Charles I in 1516 and then Charles V of the Holy Roman Empire in 1519, might best be defined instead as a series of partnerships administered through a venture capital structure. Yet the matter of origins, of when the "business of empire" begins and ends, brings us back to the matter of metaleptic narrative and how our understandings of the past inform the roles we visualize for ourselves in the present and vice versa.

Whereas Graeber's *Debt* (2014) tells a five-thousand-year-old story of historical agents motivated to action by their debt (like the stowaway Balboa), Jason W. Moore's *Capitalism in the Web of Life* (2015) compels contemporary readers to combat climate change by changing the narrative about the origins of the predicament of the global commons. According to Moore, this begins with telling the "true origins" of climate change in what Fernand Braudel (1973) called "the long sixteenth century," beginning with the Portuguese expeditions to Africa and later Asia from the mid-fifteenth century on. For Moore, and other subscribers to the term "Capitalocene," the *double internality* of capitalism in nature and nature in capitalism was wrought on a global scale by the expansion of the Portuguese and Spanish Empires financed by the spoils of primitive accumulation, which included the spoils of Native American and African bodies and their forced labor. By coining the term "Capitalocene" to locate the origins of the planet's destruction not in the Industrial Revolution in the late eighteenth century but rather in the long sixteenth century, and not in humanity as a whole but rather in the social relations inaugurated by capitalism with coloniality, Moore issues a call to arms by reconceptualizing the time period to which we belong.

Moore's push for discursive discontinuity bears some similarities to Las Casas's own efforts to elicit legal change through emotional appeal, combined with reperiodization. As Stephen Greenblatt (1991, 81) argued, Columbus's *Carta a Luis de Santangel* (1493) launched the conjunction of the most "resonant" legal ritual, that of possession, with the most "resonant" emotion, that of marvel. With the publication of his *Brevísima relación de la destruición de las Indias* (1552), Las Casas injected dissonance into the legal resonance between possession and the marvelous. By redefining the "marvels" (*maravillas*) of the New World as those atrocities committed against its Indigenous peoples in the name of "so-called conquests" (*las dichas conquistas*) ever since their discovery in 1492, Las Casas effectively made an argument about what was truly, *really* marvelous about Spanish illegal possession in the Indies. Not to put too fine a point on it, the act of naming, one of those edenic speech acts imbricated with the cultivation of the law, becomes repurposed for the production of counternarratives aimed at subverting the rationale encased in the use of a term or a phrase. It is a radical form of nominalism employed in the belief that a change of nomenclature will, in turn, inspire a change of consciousness and thus comportment.

As we shall see in the following chapter, by prohibiting the use of the terms "conquista" and "conquistador" and encouraging the use of synonyms such as "discovery" or "explorer" instead, the Spanish *Ordenanzas* of 1573 emptied Las Casas's conscientious use of language of its radical possibility. At the same time, while I understand Todorov's contention that the prohibition of the word *conquest* but not the activities comprised thereof exemplifies the cynical deployment of euphemism in the defense of empire, I argue instead that the prohibition responds to an underlying anxiety about the economies of moral and material values deployed by the conquest on a global scale. The *Ordenanzas* attempt to regulate, not obfuscate, the force of Spanish subjects' appetite to expand the reach of the Spanish Empire, the worst of which, by 1573, had made the invocation of "conquista" cringeworthy among its defenders as well as its detractors.

Another aspect of modern subjectivity has been attributed to the widely published *Carta a Luis de Santangel* by Christopher Columbus. In mapping how Petrarch's language of desire (cupiditas) shaped the colonization of America, Roland Greene (1999) has contributed to the difficult task of tracing the origins of the Spanish Empire's unique subjectivity, beginning with the *Rime Sparse* as refracted through Columbus. In the spirit of the radical nominalists, however, I would note that while *conquistar*, "to conquer," may still be used as a synonym for wooing the beloved in Spanish, the sedimentary strata of the problematic phrasing are, more often than not, understood by most speakers. We might speak now of a "love conquest" (*conquista de amores*), but we do so apologetically, in quotation marks, conscious of the reverberations of the phrase in our shared history.

I would propose instead that the most pernicious trope of venture capital remains the synonymous use of "caritas" (grace, charity) and "cupiditas." Working along the paradigmatic axis of utterance but also of silence, its figures have become so ingrained as to become habitus, defined by Pierre Bourdieu (1990, 56) as "embodied history, internalized as second nature and so forgotten as history; [it] is the active presence of the whole past of which it is the product."[10] Habitus, therefore, is the synecdoche of history as lived. In *The Civilizing Process* (2000), Elias referred to the habitus—*hexis* (state of being) for the Greeks—of European polite society as a "second nature" that was produced by a transformation over the *longue durée* of modernity and increasing thresholds of shame and repugnance, eventually comprising all forms of comportment. While Elias implicitly accepted the "constructedness" of habitus, Bourdieu's use of the term explicitly acknowledges this artifice that is, nonetheless, experienced as "second nature." My own concern for the metaleptic habitus of venture capital, however, does not eschew the possibility of subjectivity, a view suggested by Bourdieu in his reflection on the sources of historical action.[11] By the mid-sixteenth century, the synonymous use of "charity," "grace," and "cupidity" was embodied as a metaleptic habitus. This was quite a feat, considering that in becoming so embodied, in order to become "second nature," the metaleptic habitus of venture capital had to override the older ingrained tropes surrounding usury, which represented capital breeding capital as an *unnatural* growth.

CALL US INDIANS, Guamán Poma demands of Felipe III, because we are closer to God. The indios, he contends, embody their proximity to God, a relationship that is expressed in their name, *in-dios* (in-God). By figuring their relationship to the divine in terms of metonymy and metaphor, "indio" would no longer be an identity entirely contingent on the error of one man, Christopher Columbus, and his confusion of the Indian subcontinent and their peoples for the lands and peoples that stood in his way. Instead, Guamán Poma empties the epithet of its history of errors and raises Andean topography and its peoples to the heavens. *India*, according to Guamán Poma, comes from *tierra en el día* (land, earth, or even world in daylight), and for this reason natives of that part of the world are called "indios." Rather than an erroneous name, "indio" is the perfect name for a people who are godlier than the Spanish, whom Guamán Poma represents as gold eaters (fig. 3).

Guamán Poma insists on an etymology for *indio* based on similitude and contiguity and dates it to the same moment when Santiago the Apostle would have arrived on Galicia's shores, according

Figure 3. Felipe Guamán Poma de Ayala, "Conquista. Guaina Capac Inga, Candía, español." Courtesy of the Royal Danish Library, GKS 2232 kvart: Guamán Poma, *Nueva corónica y buen gobierno* (ca. 1615), page [369 [371]].

to local Spanish lore. Taking cues from missionaries who, on seeing some similarities between native Amerindian rites and beliefs, posited an earlier evangelization in apostolic times, the true discovery of the Tahuantinsuyu—the world circumscribed by Andean thought and experience—occurred during the first evangelization, or the reign of Inca Sinchi Roca, by Saint Bartholomew, one of Christ's apostles who "salió a esta tierra y se bolvió" (came to this land and returned [to his own]) (Guamán Poma 2001, fol. 368 [370]). Via Bartholomew's missionary work, only Cuzco and Callao received Christ's good news in the first wave of global evangelization.

The first Indian encounter with Christian love was thus based entirely on caritas. The second was driven by the cupiditas of Spanish subjects who, for Guamán Poma, were nominally Christian in the sixteenth century, just as the indios themselves had fallen into apostasy under Inca rule. He reproduces the encounter between a "Spanish" conquistador and an Inca in Cuzco as a dialogue in Quechua and Spanish that takes place under the title "Conquista. Guaina Capac Inga. Español." In the drawing, the Inca Guayna Capac asks Pedro de Candía, "Do you [second-person sing.] eat this gold? *Kay qurita-chu mikhunki?*" To which Candia responds, "Yes, we eat this gold. *Sí, este oro comemos.*" In this scenario, we have a native interpellation of the conquistador and an answer, given in Spanish, that reflects the emphasis made in confession on extirpations of idolatry. Guamán Poma was familiar with this mode of examining conscience, having served Friars Martin de Murúa, Cristobal de Molina, and Cristobal Albornoz in their campaigns in his native Huamanga in southern Peru. He would have learned that true conversion, following Augustine and the Dominican order's practice and theory of conversion in the Americas, is only possible once you deny your past and make it past, what Edmundo O'Gorman (2006) called a process of "self-annihilation." And yet in Candia's utterance, we have confession but not repentance.

Guamán Poma's drawing inverts the positions between confessor and penitent. The Inca's question, "Do *you* eat this gold?," elicits a response in Candia that reflects on the conquistador as a class of people under Spanish law: "Yes, *we* eat this gold" (emphasis mine). The choice of Candia, who was Greek, to represent the category "Spanish" in the drawing brings the connection between national law and international commerce in Spanish imperial expansion to the fore; like the Portuguese Magellan, or the German Fuggers in Venezuela, Candia represents the *Spanish* in the Spanish Conquest to the extent that his conduct was regulated by the contracts signed in Seville and the laws promulgated by the Spanish monarch and the Spanish Council of the Indies. In this foundational scene of conquest of the Tahuantinsuyu, the "Spanish" individual's confession is not only damning to himself, but to an entire group of people, encompassed by the "we" of his answer, performing a kind of ethno-suicide. Candia's confession under the Inca's gaze reinvents the cannibal of Columbus beneath the authorial vision of an Indian Christian. Eating a eucharist of gold, the conquistador, or empresario—the entrepreneur, in short—confesses to idolatry but fails to repent.

Guamán Poma's narrative of conquest makes an appeal to the Spanish sovereign's conscience. In the *Nueva corónica*, Candia's return to Spain sets off a rumor of gold, fueling cupidity, which in turn produces dreams, quasi-nightmares, and riots (*alborotos*). He proposes a re-volution in peninsular consciousness, a great reversal of time, space, and collective wills. The Spanish Conquest by another name would be *pachakuti*, the trope in Andean narratives for a cataclysmic renewal of time and space: in other words, in another world, the Spanish conquest brought about a pachakuti in the land of Castile.[12] Significantly, Guamán Poma never returns to the providential arc of the original mission, Saint Bartholomew's apostolic endeavor; instead of providence, these new voyages were fueled by the unruliness of adventurers and idolaters.

As told by Guamán Poma, the moral and ethical upheavals provoked in the inhabitants of the Iberian Peninsula by the (good) news of Peruvian gold resonates with the assertion made by Anne McClintock (1995, 7) for Victorian England that "imperialism is not something that happened elsewhere." Indeed, Spain is the "elsewhere" to Guamán Poma's India, whose tall mountains bathed in sunlight keep the lands and their peoples closer to God's good graces. In the shadows of the Indies, the Spanish were completely transformed by the rumors of gold and riches to be had; they could not eat or drink or sleep as their thoughts were consumed with lust for the treasures across the Atlantic. These symptoms, not coincidentally, are those associated with "love sickness," as the lover is consumed by thoughts of his beloved who, often, he has never even seen. What Guamán Poma describes is a collective awakening of appetites and purpose, a change in consciousness so drastic as to galvanize the mobilization of life and capital in pursuit of great wealth beyond the western horizon.

As gold eaters, the Spanish in Guamán Poma's rendition of events in the sixteenth century takes Spanish cupidity to task as an *unnatural* appetite but also parodies Spanish visions of Indigenous monstrosity that had taken form with the "discovery" (the "factish," again) of Columbus's cannibals, constituted with the publication of the *Carta a Luis de Santangel* and its various translations thereafter into Latin and European vernaculars.[13] In effect, Guamán Poma's narrative of the genesis of the gold eaters, condensed in the image in figure 3, offers not only a chastening rebuke to his royal interlocutor, Felipe III, but also an opportunity to gain self-knowledge through the proverbial looking-glass: you and your world have not been the same since your encounter with us. Moreover, his scathing representation of Spanish cupidity mines the edifice of "love interest" in favor of a Thomist condemnation of usury and cupidity while reviving the tensions between heterogeneity and homogeneity in questions surrounding language, the Eucharist, world dwelling, and money.

Guamán Poma's redefinition of *indio* from the Andean highlands implicates the global reach of Catholicism, as the term had become synonymous with "native" to most Indigenous peoples (*naturales*) under the yoke of the Christian Spanish Empire, with the exception of those peoples "native" to the Iberian Peninsula. As shown by the recent work of Nancy Van Deusen (2015) on the trials of indios and indias to receive their freedom throughout the sixteenth century, "indio" was a descriptor that conflated peoples from the East and West Indies, China, the Moluccas, India, Brazil, Hispaniola, Mexico, and Peru. It was a term, per Owensby after O'Gorman, used by Iberians to define a new "people" in a relationship defined—a priori—as unequal between the pagan periphery and the Christian metropole. When unraveling the irony of the indio explored from Guamán Poma's perch, his material understanding of an identity—construed during the sixteenth century to apply globally to "local" peoples—reveals the fault lines of universals and the a priori limits of jus gentium doctrine. The invocation of jus gentium only makes sense in an imperial context; that is, it depends on a distinction between a "local" norm and the "supralocal" context, such as the Roman Empire whose laws and practices of conquest bequeathed to posterity the terminology "laws of peoples."

As long as the customs of colonized or to-be-colonized peoples did not conflict with universal law, imperial magistrates were to respect local practices. Yet, as Clarke has observed in *Fictions of Justice* (2009), in her analysis of the application of international human rights law in Africa today, many "laws of peoples" themselves aspire to universality.[14] This confrontation of worldviews, and the laws they uphold, is a leitmotiv in the most memorable accounts of conquest, often because such an encounter exposes the narrator—in the *relaciones* genre, the conquistador—to another habitus and, thus, another system for tabulating moral, symbolic, and material values that is difficult to reconcile with his own. In those moments, the providential arc of conquest foretold is interrupted, and alternative narratives may be contemplated if only to be debunked through various discursive strategies. An anecdote from the *Segunda carta de relación* by Hernán Cortés exemplifies the unexpected contingencies encountered by a conquistador in dialogue with Indigenous peoples, scrambling to find the appropriate response to questions that undermined the Spanish Empire's legitimacy.

In the encounter between Otlintec and Cortés, reported by the latter, we can hear traces of the oxymoron "local cosmovision" and jus gentium, understood in the performative, metaleptic understanding of those "local laws of as-yet-to-be-conquered peoples" and the refusal of a subject of the latter category to understand, let alone comply with, this interpretation.[15] Early in the letter, Cortés elucidates the paradox of his enterprise by reporting a local native lord's difficulty accepting the world vision outlined in the *Requerimiento*, a scripted performance for Indigenous subject- or slave-making used by every conquistador between 1513 and 1542. Through the rhetorical device of indirect speech, Cortés conveyed to Charles V—Holy Roman Emperor, king of Spain—Otlintec's incredulity when he is asked to consider the existence of another lord and lands beyond Mohtecuçoma's reach following the reading of the *Requerimiento* by Cortés:

Le pregunté si él era vasallo de Muteeçuma o si era de otra parcialidad alguna, el cual, casi admirado, de lo que le preguntaba me respondió diciendo que quién no era vasallo de Muteeçuma, queriendo decir que allí era señor del mundo. (Cortés 1993, 171)

[I asked him if he was a subject of Mohtecuçoma or of another realm, and he, almost startled by what I had asked him, responded by asking, who wasn't Mohtecuçcoma's subject, meaning to say that over there he was the world's overlord.]

Confronted with the native lord's refusal to submit to the Spanish sovereign or give him "some gold," as begged of him by Cortés, without Mohtecuçoma's consent, the clash between worldviews and rival sovereigns, via their subordinates, also offers Cortés an insight into another's "local" cosmovision as a rebuttal to his own, equally local cosmovision. At the same time, Otlintec's rational refusal to imperial Christianity plants the seed for Cortés to construct a narrative of conquest that favors indirect rule of Indigenous civilization and its productivity via the native sovereign's consent given to a foreign sovereign. In this passage, Cortés concludes the exchange with Otlintec by portraying his own pragmatism and dissimulation as evidence for his good governance to his own overlord, Charles V. In this case, Cortés eschews enforcing the submission or slavery/death option given by the *Requerimiento* to new Indigenous subjects of the Spanish crown in favor of dissembling and delay.

Had Cortés signed a *capitulación* (contract) with the Spanish sovereign before embarking on his conquest of Mexico, he would have had contingency plans drawn up, itemized, and accounted for in order to maximize corporate profits, much like those signed earlier between Pedrarias Dávila and Fernando II of Aragon or later between Francisco Pizarro and Charles V. As noted earlier, Cortés and the men who followed him were motivated by debt, which often set off a spiral of profitable violence (Graeber 2014, 313–19). The drive to acquire the booty and the manpower to accumulate enough wealth to settle accounts often left these empresarios in the red and seeking new ventures and further indebting themselves, a pernicious cycle.

Cortés may not have had the royal charter to conquer Mexico, but his letters to Charles V insisted that his actions were compliant with common practices, as if his contingency plans had been drawn up in the Casa de Contratación in Seville. Where Cortés did *not* follow "proper" procedure, as in the massacre of Cholula, he justified his decision for preemptive action (*mejor prevenir que ser prevenido*) by conjuring up the figure of a well-functioning marketplace whose activity ignores the specters of the recently killed native subjects. According to Cortés, the loss of lives, at least three thousand men by his own count, was not felt three weeks after the massacre because "quedó la ciudad y tierra tan pacífica y tan poblada que parescía que nadie faltaba della, y sus mercados y tratos por la cibdad como antes los tenían" (the city and the land became so peaceful and populous that it seemed as if nobody was missing, and the markets and commerce in the city worked as they had before) (Cortés 1993, 194). The counterfactual as employed by Cortés—the marketplace continues to function *as if* none were missing—will be reworked by Bartolomé de Las Casas and other missionaries to place the onus of contingency back on the immortal life of each agent of empire.

To counter the trading fictions employed by Cortés and other entrepreneurs in profitable violence like him, Christian missionaries and thinkers such as fray Antonio de Montesinos, fray Bartolomé de Las Casas, and Archbishop Jerónimo de Loaysa raised the specters of Indigenous death over the conscience of any Christian who may have profited from the de facto Spanish Empire in America. By limiting access to the sacraments of penance, and therefore communion, as an impediment to circulation, these religious imposed and enforced de facto excommunication from below. The refrain "No hay remedio" (There is no remedy) for the death and destruction of the conquest reinforced the material conditions for barring absolution to subjects who had benefited from the conquest. And yet as the midcentury mark passed, how the conquest was told by Christian activists such as Las Casas soon implicated Iberian colonial societies in their entirety. Soon the incommensurability of conquest's destruction became synonymous with the impossibility of materially reversing the state of emergency even as it was considered morally untenable.

Las Casas narrates the story of conquest in *De Thesauris* as an ac-cumulation of moral and material indebtedness accrued by the con-quistadors (and, by extension, the crown in receipt of the *quinta real*, or royal fifth) with no feasible way to settle accounts with the Indige-nous peoples of America. With "no remedy" in sight by the end of his life, Las Casas could only conclude that an empire perpetually in hock to its new, Indigenous subjects for the sins of illegal invasion would never be able to settle accounts with them and, therefore, must return to the Iberian Peninsula. In his campaign for impossible remedies, Las Casas narrated iterations of the origins of the injustices suffered by In-dians of the Americas, Africans, and Indians of the subcontinent in his *Historia de las Indias* (1527–61). In *Habitations of Modernity* (2002, 31), Dipesh Chakrabarty contended that origins—especially violent origins that give way to modernization processes—lure the intellectual into redacting their narration, especially when the intellectual feels im-plicated in the legal conservation, through state powers, of that foun-dational violence. From his *Memorial de remedios para las Indias* of 1516 to *De regia potestate*, published posthumously in Frankfurt in 1571, Las Casas signaled one or another event, including those of 1492 but also before, as the origin of systematic oppression in the Indies, which he mapped on a global scale—from the Canary Islands to Goa— in the *Historia de las Indias*, written over more than three decades.

The multiplicity of events described by Las Casas as "origins" of foundational violence expose turning points where other decisions and actions could have led to other outcomes and offer a counternarra-tive to the dulling effects of the bureaucratic machine whose purpose, as Weber (2009a, 196–244) points out, is to ensure both the perpetu-ation of the enterprise, state or otherwise, and the homogenization of outcomes beyond the discrete identities of individual agents. Indeed, identity in bureaucracy—whether to administer the state or the enter-prise, and Weber reminds us that they are intertwined, especially in the beginning—recalls its etymology from *item* in Latin, meaning "the same." The entangled relationships between individual entrepreneurs and imperial bureaucracy complicated moral and material account-ability at a time when double-entry bookkeeping and the sacrament of confession were on the rise. If sin and debt had become largely syn-onymous by the turn of the seventeenth century, as Graeber (2014) has argued, the Christian activists and Indigenous leaders who questioned, undermined, and also used and participated in imperial bureaucracy to reallocate moral credits and debits offered alternative visions for collective governance beyond the homogenization of outcomes motivated by the pursuit of profit and neophytes. Many went unrealized, or are considered failures, but in offering alternative projects their examples serve to denaturalize—then and now—the metaleptic habitus of venture capital in the conquests and its teleology that reiterates the inevitability of its success on a global scale.

It is a well-known fact that the Spanish monarchs did not make majoritarian capital investments in their overseas ventures. This limited or no capital investment, coupled with the 20 percent (quinta real) received by the crown in each venture, was and has been a cause for consternation among rebellious *encomenderos* (landholders) as well as theorists of imperial power from the sixteenth century to the present. The Spanish structure for financing what Kamen (2004) has called the "business of empire" has led Grafe and Irigoin (2012) to argue that the Spanish Empire was nominally an absolutist state that constantly had to negotiate with various stakeholders in order to endure. Rather than distinguish between state and enterprise, however, I contend that these public-private partnerships *are* imperial, or, rather, that empire is produced *by* this entanglement, not *despite* it; the role of British and Dutch trading companies in expanding empires in Asia is well established. The Iberian empires, which precede their northern European rivals, also employed this model.

The question remains, however, what is the dispositif that allowed the Spanish crown to exert enough power to expect the quinta real from its entrepreneurs in profitable violence while sustaining negligible capital investment in these ventures throughout the sixteenth century. Put another way, it is remarkable that more conquistadors and encomenderos did not raise arms against the Spanish crown once it became apparent that the ongoing payment of the quinta real had made it a majoritarian stakeholder in ventures where they had not risked pieces of eight, a horse, or a ship. The same could be asked today of young entrepreneurs in Silicon Valley who take their technological innovations to venture capital firms such as Bain Capital or Benchmark Capital when the firms themselves do not actually invest capital but expect "carried interest"—generally set at 20 percent, similar to the quinta—after the liquidation of each fund. After several rounds of financing, the venture capital firms often end up with the most equity in these ventures. Casting their lots with the venture capitalists, the entrepreneurs make the gamble that there is more to be gained from loss of equity: rapid scalability. Graeber (2014) has presented the debt spiral of entrepreneurs as the motivating factor behind schemes, such as that of Cortés, to risk everything—including the destruction of ships—to win the jackpot, even if it meant disobeying crown officials at first.

The dispositif for Christian *and* capitalist empire formation is venture capital with a long-scale or general partnership between the church and the crown and short-scale or limited partnerships between entrepreneurs and investors. Recurring to a term coined in the twentieth century may seem anachronistic, yet conceptual gaps arise when scholars maintain the categories used by economic and imperial powers to obfuscate the interrelations of the state and private actors when analyzing violence and its origins in colonialism. In the case of colonial India, Barkawi (2010) has argued against dating the origin of British crown rule to the Government of India 1858 Act in order to better conceptualize the continuity between the power relations introduced by the British East India Company, with its private armies, centuries earlier: "The choice of term [private violence] already suggests that organizing force beyond the jurisdiction of the local state is abnormal. It means literally beyond the jurisdiction of the local state, indicative of the juridical character of much of the reasoning behind employments of Weber's definition of the state. A gap is opened between juridical and de facto relations, a gap one could drive an army through, but an army opaque to social scientific inquiry based on juridical premises" (Barkawi 2010, 37). When entrepreneurial violence is reconceptualized as the norm, not as the exception, in the imperial state formations of globalized capitalism, our choice of terminology as scholars must reflect this change in approach in order to visualize the opaque wedge of public-private and crown-state partnerships, the kind that Guamán Poma allegorized on the same boat. I show that financing and management practices in partnerships for profit over centuries have followed the venture capital structure under different names. In the sixteenth century, the partnership of crown, church, and entrepreneurs in profitable violence was known as a conquista, whose successes were accounted for materially and morally with a push to expand new subjects for trade and labor, on the one hand, and Catholic neophytes, on the other.

I contend that the Portuguese and Spanish modi operandi for imperial business are obversed in the venture capital dispositif. Whereas the Portuguese monarchs were major stakeholders in each venture and were charged 20 percent by their admirals on the liquidation of each fund (*a quinta do admiral*), the Spanish monarchs operated conversely and thus received the royal fifth, or quinta real. The lack of direct crown investments on the Hispanic side of the Iberian Peninsula has been noted by scholars such as Arrighi, Braudel, D'Arienzo, and others before me, in contrast to the Portuguese style of "monarchic capitalism," a term coined by Nunes Dias (1963) in his eponymous book to refer to Lusitanian imperial enterprise and the Portuguese monarchy's majority stakeholding in each venture.

The relationship between capital investment, monarchical power, and imperial scalability will thus drive this inquiry as I seek to map how the discourse of love informed and was informed by the structures of venture capital in the conquest. Indeed, in *De procuranda indorum salute*, José de Acosta lamented the stark contrast between Portuguese and Spanish modi operandi for imperial expansion. For Acosta, direct and majority investment in each conquest would have guaranteed greater power of the sovereign from the metropole over subjects, old and new alike. However, as already observed by Kamen, the majoritarian stakeholding by the Portuguese monarchs limited territorial expansion, at least initially, to the archipelago-like *feitorias* (trading posts) dotting the shores of Guinea (western Africa) since the mid-fifteenth century and Brazil in the sixteenth century. These connections and differences between the Lusophone and Hispanic empires are mapped out in the first four chapters, whereas chapter 5 explores the paradoxes in the metaleptic habitus of venture capital created and exposed by Indigenous elites of the Andes as they negotiated with Felipe II to purchase their local sovereignty from Spain through the advocacy work of two Dominican friars, Bartolomé de Las Casas and Domingo de Santo Tomás.

CHAPTER 1 OPENS with a close reading of the 1573 *Ordenanzas*, which sought to erase the word *conquest* from the imperial lexicon. I also examine the contrast that José de Acosta made between the modi operandi of the Spanish and Portuguese imperial enterprises in *De procuranda*. I show how writers in the sixteenth century were aware of and anxious about the Spanish monarchs' low capitalization of these ventures and their belief that lesser monarchic investment would translate into lesser leverage, or power, over the conquistadors. With references to studies of economic historians of the Italian banking presence in Portugal and Spain, such as Fernand Braudel and Luisa D'Arienzo, and to classic economic theorists, such as Joseph Schumpeter and Max Weber, I argue that the Spanish method of imperial enterprise follows a venture capital structure.

Though known by other names over centuries, venture capital was learned from Genovese and Florentine merchant seafarers on the Iberian Peninsula. The general partners, like their modern counterparts, were the Spanish crown and the church, neither of which gave the majority of capital or labor for each enterprise; limited partners were the conquistadors, family members, and investors who contributed to each expedition in labor and in kind. I conclude that the moral and ethical exceptions enjoyed by seafaring expeditions to charges of usury and unjust war drove a wedge through the laws of nations (jus gentium) whose right to self-rule had always been exposed to the following loopholes: (1) the Christians' right to free movement for missionary work; and (2) the merchants' right to free movement for trade. The exploitation of these loopholes, in turn, structures the accounts (bookkeeping but also narratives) kept by merchants, conquistadors, and future historians, such as Gonzalo Fernández de Oviedo (1478–1557), so that the logic of venture capital—which offers ownership in a venture that is, as yet, unformed and where causes are confused for effects and effects for causes—takes possession of Hispanic narratives in the "long sixteenth century."

That the Laws of the Indies provided the framework for colonial narratives is largely undisputed in the field of colonial Latin American studies. Yet in the second chapter I show that commercial maritime practices largely prefigure the Laws of the Indies, which were promulgated to protect the Indigenous peoples. These practices codified in contracts (*capitulaciones*) even contradict the laws at times. In this chapter, I analyze the contracts signed between the general and limited partners in each venture, that is, between crown and conquistador, as prime examples of the metaleptic habitus of venture capital. In this way, I underscore the tension between providential design and contingency plans in the contracts signed between conquistadors (Pedrarias Dávila, Francisco Pizarro) and the crown (Fernando II of Aragon, Charles V); between the three sovereigns (King João I of Portugal, Isabel of Castile, Fernando II of Aragon, and Pope Alexander VI) in the Treaty of Tordesillas (1494); and, finally, the *Requerimiento* and the laws that governed the administration of Indigenous peoples (1512, 1526, 1542, 1543, and 1573).

Chapter 3 juxtaposes two "minor" works—or a minor episode within a larger work, as in the case of the *Historia*—by two major writers who are foundational in the field of Latin American colonial letters in order to visualize the differences between writing imperial enterprise in a metaphoric and a metonymic vein. If chapter 2 discusses the futurity of instructions, outlined in contracts, laws, a script (the *Requerimiento*), and a treaty (Tordesillas), this chapter examines two narratives that alternatively account for or erase the tensions between contingencies, on the one hand, and providence, on the other, by two *letrados*, writing officials whose place in colonial society Angel Rama (1996, 22) once described as follows: "As servants of power in one sense, the *letrados* became masters of power, in another."

I contend that islands, and the narrative archipelagoes subsumed or traced by these letrados, Fernández de Oviedo and Las Casas, respectively, serve both authors to construe their place as moral actors within the larger arc of imperial history. Taking my cue from Martínez-San Miguel in "Colonial and Mexican Archipelagoes" (2017) and Pugh in "Island Movements" (2013), the archipelagic turn of the third chapter traces periphrasis—rhetorical circumnavigation—to analyze the assemblages that give rise to islands' coloniality but also and perhaps more important for the potential of archipelagic thinking, which is inherently relational, for imagining other futures from distinct junctures in the past. I show how Las Casas ties these junctures to the island-hopping of the conquistadors, thus giving a spatial dimension to the counterfactual moments in his narrative of conquests.

The legacy of the Lascasian counterfactual made it impossible to narrate the providential nature of the conquest without examining its contingencies. Such an approach is followed by Acosta in *De procu-randa*, though, as Ivonne del Valle has argued in "José de Acosta, Violence and Rhetoric" (2012), the Jesuit missionary and scholar does not place contingency and providence in conversation. Instead, the injustices of the conquest, the conquest itself, is buried as a *thing* of the past, while the future—the reform of the Spanish Empire—remains the only realm for moral discernment in Acosta's treatment of the Spanish Empire. Other scholars of Acosta and Las Casas, such as Anthony Pagden, Rolena Adorno, and José Cárdenas Bunsen, have placed these two authors in conversation before me, especially in regard to the similarity in their comparatist approach to ethnology. However, in the fourth chapter, I draw a contrast between the two writers for their treatment of the related concepts of freedom and salvation and their views on mercantilism as models for missionary work. Though Acosta never mentions Las Casas or his works by name in the *Historia natural y moral de las Indias* (1590) and *De procuranda*, I show that Acosta is in fact polemicizing with the body of Las Casas's thought, especially with the Dominican friar's *De unico vocationis modo* (ca. 1538),[16] *De Thesauris* (1563), and *Doce dudas* (1564). While showing that the radical theology of Las Casas haunted Acosta, I also bring to the fore the continued haunting of modernity, whose indebtedness to the grave and labor theft of the Indigenous peoples of America remains largely unrecognized.

Finally, chapter 5 discusses the paradoxical bid of the Indigenous elite, *curacas*, to purchase their inalienable sovereignty from the Spanish monarch in 1560. While their failed bid exemplifies the contradictions of the Spanish-Christian imperial enterprise, it also highlights the agency of Indigenous elites of the Andes in responding in kind, with their own trading fictions. In this way, this chapter intervenes in current polemics about the agency of the Indigenous in their negotiations with their invaders while resisting any conflation between Indigenous elites, who were exempt from giving tribute in labor and in kind, and the indios, in whose name they ostensibly spoke. I show how the negotiating document redacted by the curacas of the Mantaro Valley, and presented by Bartolomé de Las Casas and Domingo de Santo Tomás to Felipe II, exposed the false equivalence between greed and

Christian charity, which had become acceptable in colonial and evangelization discourse by the mid-sixteenth century.

THE DEATH OF Bartolomé de Las Casas in 1566 marks its own event
in the standoff over the Spanish conscience in the conquest. His death
coincides with the emergence of probabilism in the schools of Salamanca and Alcalá de Henares, and it was probabilism that placed the
enforcement power of absolution in peril.[17] For fifty years, beginning
with fray Antonio de Montesino's cries in the moral wilderness of Hispaniola in 1510, his invocation of Saint John the Baptist inspired generations of religious to refuse absolution to anyone who had profited
from the conquest. Their boycott began with the conquistadors and the
holders of *encomiendas* (leases of labor and tribute from Indigenous
communities), but by the 1560s, Archbishop Loaysa and Friars Bartolomé de Vega and Las Casas had concluded that hardly none could
claim *not* to have benefited in some way from the conquest, whose il-
legality, for them, was no longer in doubt.

Yet with probabilism, the state of doubting whether an action
under consideration is sinful or not no longer imposed a roadblock to
taking that action. In other words, if a Christian subject is in doubt
about the lawfulness or unlawfulness of an action, it is permissible (i.e.,
not sinful) to follow a solidly probable opinion in favor of liberty even
though the opposing view is more probable. Thus, the probabilist position, favored by Francisco Suárez in the *Disputationes* and defined by
Bartolomé Medina in his "Expositio 1am 2ae S. Thomae," placed the
enforcement power of absolution in peril. Under probabilism, the boycott against an entire class of people, such as the holders of encomienda,
or conquistadors, that had been exercised by Las Casas, Loaysa, and
Vega when they refused them penance, and therefore communion, lost
its immanent force. No longer would the moral imperium of their
boycott serve as a force for counter-imperium to the Spanish *merum
imperium* (exercise of material power) in the Indies. And yet their
collective doubt—though seemingly resolved by the pragmatism espoused by probabilism—transcended whatever respite Felipe II had
found from his doubts over his sovereignty in the Indies, doubts his
father, Charles V, had never resolved but enacted into law on two sepa-

rate occasions, in 1526 and 1542. The transcendence of the Lascasian position may be seen in the invocations of his thought, and phrasing, by liberation theologists such as Gustavo Gutiérrez and Indigenous theologists such as Eleazar López Hernández. And when practices such as the boycott or the roadblock take a stand against capital's circulation, his radical position becomes immanent as his thought and practice continue to haunt modernity.

CHAPTER ONE

On the Same Boat

Iberian Ventures in Christian Conquest

Too much disorder leads to order. A saying meaning that superfluous expenses and extravagance lead to poverty and misfortune;
and this requires moderation and good government.
—*Diccionario de Autoridades*

Vitoria's work demonstrates, for instance, the centrality of
commerce to international law, and how commercial exploitation
necessitates war.
—Antony Anghie, "The Evolution of International Law"

What's in a conquest? The *Ordenanzas de descubrimiento, nueva población y pacificación de las Indias dadas por Felipe II, el 13 de julio de
1573, en el bosque de Segovia* (in)famously proscribed the use of the
word *conquista* (conquest) and called instead for the use of *descubrimiento* (discovery) or *pacificación* (pacification): "pues hauiendose
de hazer con tanta paz y caridad como deseamos no queremos que el
nombre dé ocación ni color para que se pueda hazer fuerça ni agrauio
a los Indios" (for [as this activity is] to be done with as much peace and
charity as we so desire, we do not wish for the name to give occasion
for the use of force or injury against the Indians) (Morales Padrón
2008, 495). Recalling Juliet's plaintive question, we might ask, like
Tzvetan Todorov (1984, 173) in his reading of the twenty-ninth ordi-
nance: What's in a name? Surely, it is only the word *conquista* that was
banished and not the activities comprised thereof.

Throughout the sixteenth century, Spanish law had made explicit the conflicts of interest inherent in ventures that pursued enterprises for both moral and material gains. Yet by that century's close, the changes in discourse brought about by Spanish empire building had also unleashed a new subjectivity with formidable potential, capable of reconciling paradoxes and marrying antitheses. Love and interest could be yoked in the service of empire seemingly without conflict, though the use of the word *conquista* would become problematic, an unwelcome reminder of the embattled positions and conflicts of interest that it signified. Thus, according to the authors of the *Ordenanzas* of 1573, the word *conquista* had impelled the Spanish crown's subjects and agents to act in ways that contradicted the crown's desired objectives, of both a material and a religious order, and its love for its new subjects in the New World, the Indians.

The rationale for the twenty-ninth article of the *Ordenanzas* seemingly argues in favor of a correspondence between the *name* for violence (conquest, discovery, and pacification) and the actions performed under its aegis: "pues *haviendose de hazer* con tanta paz y caridad como *deseamos* no *queremos* que el nombre dé ocación ni color para que *se pueda hazer* fuerça ni agrauio a los Indios." The passage offers a striking contrast in subject positions between the active "royal we" (*deseamos*, *queremos*) and the impersonal construction for both prescribed and proscribed actions (*haviendose de hazer, se pueda hazer*). The law's circumlocution ironically delineates yet another island to be discovered, populated by the very people, formerly known as conquistadors, who seemed to be (un)doing the bidding of the sovereign. Yet empire would gloss over their agency while alluding to the wrongs (*fuerça, agrauio*) committed by these agents, placed in parenthesis by the letter of the law.

In the classical trope on language and civilization, grammar plows the formerly sylvan fields and shares its function with the *nomos*, the rule of law that lays claim to an ordering of the world and the right to uphold it by violent means.[1] In 1573, less than one hundred years after Antonio de Nebrija (1931, 1) made his claim that language has always been the handmaiden of empire, the laws of Spain would tame unruly subjects by offering a change in nomenclature and a law whose grammar strangely reproduced the unruliness it claimed to disallow. Yet the question remains, Was this name change just a lexical sleight of hand?

Nominalism at its worst? If so, who was fooling whom? And why would *conquista* serve as an excuse for unruly behavior? Also, why had it become antithetical to the "new" mode of imperial expansion?

At the heart of conquista and its discontents, the knotty matter of subjectivity, agency, and stakeholding comes to the fore. After all, putting cynicism to one side, the *Ordenanzas* of 1573 espouse the idea if not the belief that removal of a word—*conquista*—could change the behavior of the laws' agents. Thus the new laws' premise for proper functioning implicitly envisioned a top-down hierarchy in which the comportment of an unruly mass could be dictated by the letter of the law.[2] In this imagined scenario, the sovereign in the metropole imposed his vision of order on the periphery by taming his old subjects (former and would-be conquistadors) for the benefit of his new subjects (Indians) through an agentless process of "pacification" and "discovery."[3] Yet empire building had been a collective endeavor from the start, and its ownership was an ongoing matter of contention and negotiation.

Less than a century after Christopher Columbus sailed southwest from the Port of Palos in 1492, the Spanish Empire sought to turn over a new leaf in its scripting of violence for material and spiritual gain. So what was at stake in the name change? Who were the stakeholders in this activity (conquista no longer) in the first place? A reading both in and against the spirit of Machiavelli, who had praised the appearances of religiosity employed by Ferdinand II of Aragon in various wars as instances of "cruel piety" in his 1532 work, *The Prince*, might criticize the laws as yet another example of a Spanish monarch's cynicism, though performed this time by his great-grandson, Philip II. And yet beyond a critique of the facade of Christian piety in the *Ordenanzas* of 1573, I contend that they both mask and unmask the moral and material vulnerabilities of conquistas as perceived by the sovereign: the corporate enterprise of Spanish imperialism and its dependence on individual risk takers, merchants, conquistadors, sailors, and bankers (often embodied by the same person), all of whom held a stake in this profitable violence. Imperial enterprise was an aggregate of private undertakings by a series of conquistadors for profit in partnership with the Iberian crown and the church over the long sixteenth century (per Braudel [1973], ca. 1450–1650).

Yet the question of how to conceptualize the relationship between "private undertakings" and "imperial enterprise" remains. In 1573, the

Ordenanzas' heavy-handed attempts at erasure only served to underscore the double-sided coin of the Spanish empresa, which in its original sense comprised many "enterprises," military, commercial, and symbolic. A little over a decade later, in 1589, it was the Jesuit scholar, missionary, and inquisitor José de Acosta, who succinctly described the problem of the Spanish Empire in terms of capital when he published *De procuranda* in Seville. Acosta took for granted that the aggregate imperial enterprise (conquista) as a historical and theological-political process had been constructed from the individual undertakings (conquistas) of each conquistador with little capital, material, or labor investment by the sovereign. In the early modern period, venture capital was known and practiced under a different name: the *commenda* or, in Romanist jurisprudence, the *societas pecunia-opera* (*in qua alter imposuit pecuniam, alter operam*) (the partnership to which one contributed the money, the other, the labor) and the *contractus trinus* (triple contract). Venture capital thrived as an alternative to loans charging interest, especially from the mid-fifteenth century on (Noonan 1957, 133–53). The model Acosta describes for militant empresas in the West and East Indies is surprisingly similar in form if not in nomenclature to venture capital as it is practiced today.

VENTURE CAPITAL, SOCIETAS, AND CONQUISTA

Before turning to Acosta's critique of imperial enterprise, stakeholding, and monarchical power in the sixteenth century, it is helpful to review what is generally understood as venture capital today in order to visualize the financing and managerial structure behind a global revolution in thought and practice, one that was, nonetheless, naturalized and thus remains invisible. The venture capital structure of financing and management contributes, on the one hand, to what Graeber (2014) has described as the spiral of debt and violence and, on the other, to the reach of monarchical and pastoral power as moral underwriters of the material imperatives for global expansion. Within this paradigm, the ongoing negotiation of the terms for each individual enterprise, especially those following a venture capital structure, reinforced the corporate structure of empire and the power of the crown's and the church's imprimatur. In other words, the negotiation afforded by enterprise concomitantly provided for a new subjectivity to develop between the poles of coercion and consent.

Known as conquistas, the enterprises embarked on by the crown, the church, and the conquistadors and their crews were a series of venture capitalist schemes in which crown and church provided "managerial expertise" to conquistadors and crew in exchange for the quinta real (20 percent) and tithing (10 percent), respectively. In the twentieth and twenty-first centuries, venture capital thrives on the commercialization of science and technology. The general partners of venture capital firms raise money for and find and evaluate entrepreneurial ventures and participate in their management to increase their value as rapidly as possible, yet they do *not* provide the majority of capital invested in any one fund (Freeman 2005, 146–47). In this way, the general partners invest and distribute capital provided by others, known as the limited partners of sequential endeavors (or funds). Limited partners are often family members or acquaintances of the general partners and the entrepreneur, though more often than not the entrepreneur will make the least capital gains among all the partners of a fund, even if the enterprise was his original idea. If general partners are valued for their ability to build a corporate structure for the greater profit of all stakeholders, the entrepreneur is credited with having the original idea and bringing it to fruition "against all odds" (Schumpeter 1983, 85).

Joseph Schumpeter, the renowned Harvard sociologist, placed particular emphasis on foresight as a defining trait of the entrepreneur's character: "Here the success of everything depends upon intuition, the capacity of seeing things in a way which afterwards proves to be true, even though it cannot be established at the moment, and of grasping the essential fact, discarding the unessential, even though one can give no account of the principles by which this is done" (1983, 85). This seminal description of the entrepreneur raises the specter of irrational belief and practice; the logic of his actions only makes sense after the fact. Indeed, as a matter of narratological inquiry, Schumpeter's definition of the entrepreneur engages in metalepsis, the confusion of causes for effects or vice versa. The unique faith of the entrepreneur—unique in that only he believes in the enterprise at hand—defines him by the tautology of success, in hindsight.[4] However, in the end, the entrepreneur's intuition will receive less remuneration than the managerial expertise of the general partners, who will also own the largest stake in the enterprise by the time the fund is liquidated.

The willingness of entrepreneurs to cede the majority of their stakeholdings to the general partners seems counterintuitive since the general partners gain their "carried interest," normally 20 percent, even though their capital investments in the enterprise are minimal if any. Entrepreneurs are willing to relinquish ownership of the enterprise to venture capital firms for the latter's valuable social networks, which are necessary for raising capital; because investment by a venture capital firm of renown gives the enterprise "legitimacy" and attracts more investors, this leads to more capital investment in the original idea and greater scalability, or expansion. Also, general partners are believed to organize the labor force more efficiently and, having navigated nascent enterprises before, can apply practices and structures learned from previous experience to the current endeavor. Thus if all goes well, even though the entrepreneur loses most of the ownership of his original idea, the distribution of risk combined with a substantial capital investment will offer a greater rate of return. For all their managerial expertise, general partners receive carried interest once the assets of each fund are liquidated. Carried interest is calculated after the original investment of the limited partners has been returned, of which, usually, 20 percent belongs to the general partners. This 20 percent often makes the general partners the owners of the largest stake in the enterprise by the time the fund closes. Despite the change in nomenclature over centuries for this method of financing and entrepreneurship, it is striking that the percentage received by general partners remain largely the same: 20 percent, as carried interest, the quinta real (Spain) or the quinta do admiral (Portugal) for each conquista.

On and beyond the Iberian Peninsula before 1492, during the long sixteenth century, militarized trading companies enjoyed a long tradition of employing the venture capital model. It should not surprise us, then, that while describing the modus operandi of imperial violence in the West and East Indies, Acosta would contrast Spanish imperial expansion to that of the Portuguese in terms of each monarchy's capital investment in any given private undertaking made in the name of imperial enterprise. For his analysis, the Spanish crown's method for remunerating conquistadors through the encomienda system was a given.

The crown could not compensate them otherwise, because the conquistadors, not the crown, had made the investments (whether in labor, in capital, and/or in kind) in each private venture (conquista) even though the crown had a vested interest—morally and materially—in the larger imperial project (Conquista). For Acosta, leases of Indigenous labor and tribute (i.e., the encomienda system) in the Americas serve as a return on the conquistadors' violence when it is understood as labor for profit, and such profitable violence constitutes the theological-political economy of Conquista.[5]

> Ac prima illa de remunerandis laboribus sumptibusq[ue]; militarium hominum, ex necessitate quadam potius quàm ex voluntate, aut religione profecta fuisse videtur. Neque enim poterat Princeps, aut per quam aegrè poterat, tantos tot hominum sudores, imo verò etiam cruores dixerim, praemio pari afficere, nisi in nouo orbe illorum virtute parto, pote[n]tiam quaestumq[ue] partiretur. Nam neque isti alioqui contenti essent, & caeteris similia, aude[n]di, aggredie[n]diq[ue], cupiditas omnis extingueretur. In Lusitanica India, quod Regum Lusitanoru[m] auspicijs, & auro parta sit, potuit penes Regem totus ille dominatus sine iusta fuorum querela retineri. Nostrorum verò hominum, quonia[m] suapte ductu & re, tanta peregerunt, longè alia ratio est. Itaq[ue] ; necessitatis, vt dixi, cuiusdam fuit, vt suo quoadam iure, vt olim Israëliticae tribus distributore Iosue, terram sortirentur, permanente tamen, quod minimè obscurum est, supremo omnium penes Regem imperio. (Acosta 1589, 317–18)

> [The idea of remuneration for the work and the expenses of the *conquistadores*, was born out of necessity rather than out of desire or religious concern. For the Prince was not able, or only with greatest difficulty, to give a suitable prize for such "toil and sweat" (what that really means is "such bloodshed"), save to divide up amongst them some of the power and the income of the New World which had been won through their fortitude. They themselves would not have been satisfied with any other prize and for the others who followed it would have quenched any desire to undertake similar ventures. In the Portuguese Indies, as all was

conquered under the auspices and through the gold of their kings, all the dominion and control were able to be kept in the monarchy, without any just protest or offense to the individuals who carried out the task. But in the case of the Indies of Castile it is a different case altogether, since private enterprise played the major part. So, as I said, it was out of necessity, as in other times, like the tribes of Israel for example, where individuals obtained the land by lot, although as is quite clear, the supreme control of distribution always remained in the hand of the king.] (Acosta 1996, 1:123)

What seems like a digression on Acosta's part from the larger "question" he explores in *De procuranda indorum salute apud barbaros* (how to "save" the Indians, or barbarians) instead offers an entry point to examine the structure of capital investment in violence and its corresponding discourse in the construction of early modern empire.

Some questions elicited by Acosta are the *unique* structure of Spanish enterprise and empire vis-à-vis other imperial competitors, such as Portugal; the division of power to remunerate and compensate *past* investments (in both capital and labor) but with an eye to future spiritual profits (i.e., Indigenous Christian neophytes); violence as "labor"; and the initial and ongoing (relative) poverty of Spanish monarchs for undertaking large capital investments. Acosta's vision of the Spanish conquista is Janus-like, in the sense that his analysis reveals the foundational and ongoing relation of dependency between conquistador/ encomienda holder and Indigenous laborer, Spanish sovereign and Indigenous subject, and conquistador and Spanish sovereign. According to Acosta, the foundational form of remuneration for each conquest had structured the conditions for *future* evangelization of Indigenous neophytes in that it remunerated the conquistadors for their *past* labor ("toil" or "bloodshed") in exchange for ongoing religious tutelage. The Spanish sovereign strikes a paradoxical figure in Acosta's description of imperial power: he has the ultimate decision-making power, yet his hands are tied. In the above passage, Acosta posited a capital-based theory to explain the Spanish sovereigns' relative lack of power vis-à-vis their Portuguese competitors.[6] As portrayed by Acosta, political power was proportional to the amount of capital invested by each agent in the scheme. He therefore failed to distinguish between general and limited partnerships as outlined above.

Acosta's arguments contribute to our understanding of violence and capitalism in the conquest to the extent that he places capital risk at the crux of sovereign power, whether or not we choose to agree with the proportional correlation that he posits between capital investment and political power. Moreover, by presenting the Spanish monarchs as debtors to their unruly Spanish subjects overseas, Acosta's vision of the partnership between evangelization and imperialism necessarily offers a sixteenth-century vision of the intersection between *auctoritas* and *dignitas* because it takes the relationship between capital investment, sovereign power, and spiritual gains for granted. Recalling how in *The Kingdom and the Glory* (2011, 83–123) Agamben traces a genealogy of the theological economy of grace as political, made manifest in the ceremonial, liturgical, and acclamatory aspects that are constitutive/constituted of modern power, Acosta's crediting and debiting of Iberian sovereignty reveals not only an evangelizer's recognition of his own dependency on the conquistador, but, ultimately, on the Indigenous subject.

In the passage cited above, Acosta's reckoning with profitable violence, evangelization efforts, and sovereign power only makes sense if violence against Indigenous subjects was accounted for as labor that needed remuneration, even when the conquistadors who held encomiendas were not fulfilling their roles as Christian tutors for the Indigenous neophytes or at the very least paying tithes, which were generated from Indigenous tribute, to the preachers for their missionary work. Yet even Acosta's ideas about remuneration take for granted that risk taking for profit in a societas was a worthy venture and not usurious. In this section I analyze the conquest in terms of venture capital as practiced by the Iberian monarchies and their partners and entrepreneurs during the long sixteenth century within a longer, millenarian tradition that grappled with the relation between risk, labor, violence, and profit and viewed the business partnership (societas) as an exception to usury.

On the Iberian Peninsula in the fifteenth and sixteenth centuries, the Portuguese and Spanish sovereigns negotiated the distribution of risk to capital in distinct ways within the venture capital model. As the main creditor to the shipowners and their employees, the Portuguese crown could claim full "ownership" of the enterprise and collect interest (or the "price of peril") because it had contributed labor to the enterprise and had sponsored the voyages in full (as noted by Acosta, "Regum Lusitanorum auspicijs, et auro parta sit"). Henry the Navigator (1394–1460) is the most obvious example of Portuguese royalty laboring in overseas expeditions, hence the argument of Nunes Dias (1963, 1:70–84, 138–202, 360–400) that Portuguese imperialism was built by "monarchic capitalism." Even if we consider that Acosta's estimations of the Portuguese monarchy's capital investments in voyages it sponsored were exaggerated, his rationale for the contrast between the two Iberian empires proposes a causal relationship between capital and dominion, leading him to recommend that the Spanish monarchy *ought* to follow the Portuguese model. Eight years into the Iberian union (1580–1640), which arose from the Portuguese crisis of succession (1578) and Philip II's invasion of Portugal (1580–81), Acosta espoused his admiration for the Portuguese venture capital model of empire with the publication of *De procuranda*.

What follows from Acosta's reasoning is a commensurate relationship between capital, dominion, and power in which the Portuguese monarch's material contributions to conquest left no room for political discourse (*potuit penes Regem totus ille dominates sine iusta fuorum querela retineri*). Yet Acosta leaves an opening for disputing claims to dominion within reason. His casuistry could accommodate "legitimate protests" (*iusta querela*) if the monarchs had *not* provided all or most of the needed capital, resources, or labor for the enterprise in question. We can infer that for Acosta the financing of the Spanish Indies, which by law had not depended on monarchic capital investments since the early sixteenth century, had left such an opening for legitimate protests.

Acosta's political equation thus refers obliquely to the *encomendero* revolts in the Viceroyalties of Peru and New Spain in the mid-sixteenth century. Encomenderos, holders of encomiendas, had disputed the Spanish monarchs' right to dominion and usufruct in terms of capital and labor contributions to the conquest of the Indies. Prior to the passage of the 1542 New Laws, which outlawed Indigenous slavery and banned the granting of new encomiendas in perpetuity, the encomienda holders had received compensation for their conquests through leases of Indigenous labor and tribute in exchange for the Christian stewardship of Indigenous neophytes, who were also new subjects of the Spanish crown. Thus the encomienda system compensated the past services of the conquistadors of the original expedition (which had resulted in material and geopolitical gains) and present and future actions (the ongoing "care" for Spain's new subjects). It was the political theological system of compensation in constituted, profitable violence that extended and institutionalized the imperial reach of the private enterprises that had performed constitutive violence in the name of the Christian empire. Whether or not the holders of encomienda were in fact complying with the second half of their contractual obligations (i.e., religious and material stewardship of Indigenous subjects) was of little concern in Acosta's allusion to just or unjust quarrels with the crown. A similar comparison between investment (in labor and capital) and dominion had led the curacas, Indigenous elites of the Mantaro Valley, in conjunction with the Dominican friars Las Casas and Santo Tomás, to outbid the encomenderos' offer to buy Philip II (r. 1554–98) out of his dominion over Peru.

If the Spanish crown's material and labor contributions had been so slim, with what right could it restrict remuneration—in moneys, tribute, and labor—and at the same time continue to profit from these enterprises? Unlike the holders of encomienda, Acosta does not push his logic to its obvious conclusion. Instead, he turns to biblical authority to designate the sovereign as the ultimate decision maker on distribution (*supremo omnium penes Regem imperio*). Acosta concludes that the encomienda system had emerged as a necessity. The encomienda system was necessary, according to Acosta, because without it the cupiditas, the lust for riches, however inordinate, of men like the first conquistadors, would be extinguished, and without cupiditas there could be no more evangelization in the Americas. In other words, this lust was itself a resource in the service of conquest that had to be renewed, even though the preferred terminology to refer to this enterprise was "pacification" or "discovery" according to the *Ordenanzas* of 1573. As an affective investment, the conquistadors' cupiditas expected material returns that, in turn, fueled more desire. In this way, according to Acosta, cupiditas functions like capital in its disjunction and alienation from its original source. However, how could capital and desire provide the basis for dominion? Acosta does not analyze the dynamics between capital investment, cupiditas, and empire in greater detail. He does, however, propose a commensurate relationship between capital investment and political dominion that he himself is not quite ready to defend. Finally, Acosta also strikes a marked contrast between Portuguese and Castilian modes of financing conquest that may have been overdetermined by his belief that political dominion may be quantified as a function of capitalization in imperial enterprises.

In contrast to the Iberian monarchs' modus operandi seen above, the Portuguese monarch's relationship to seafaring entrepreneurs falls squarely within the *foenus nauticum* tradition. This similarity between ancient and early modern financing practices for seafaring enterprises was an observation later made by Nunes Dias (1963, 2:189–213) in his book on Portuguese monarchic capitalism. Beginning as a Roman practice, when employing the foenus nauticum financing model, a loan could not be charged interest unless the creditor incurred the risk of loss on the principal of the loan. Thus, following the norms of the Roman *Digesta*, a creditor making loans to shipowners could avoid the charge of usury as long as the creditor assumed the full risk for the loss or value of the goods at sea. Interest charged for the time at sea was known as the "price of peril." If, however, the shipowner's losses arose after the journey's end, then the shipowner was liable for the full amount of the loan. During the medieval period, the foenus nauticum fell out of currency as a matter of law, though in practice it continued to be employed by sea merchants, and it was tolerated as a matter of *lex mercatoria*, or common law among merchants. Indeed, the risks of loss of life and property during sea voyages also made Aquinas more amenable to the distinction between use and ownership when such risks were pooled in the societas.

Portuguese monarchs were majority stakeholders in the fifteenth- and sixteenth-century expeditions that expanded their overseas empire. In the process, the feudal station (*mayorazgo*) of the admiral in Portugal evolved out of a venture capitalist structure, one that was never fully abandoned. In the twelfth century, the Portuguese crown contracted with Genovese merchants to mount an offensive against Muslim-held dominions on the Iberian Peninsula and in northern Africa.[7] In the thirteenth and fourteenth centuries, sea merchants linked to the large commercial houses of Genoa, with large amounts of capital to spare, were engaged by Iberian monarchs for their ships, martial expertise, and shipbuilding techniques. Ships that were used for fighting were also used for trading, and the same men were often merchants or mercenaries based on circumstance (D'Arienzo 2003, 12–59).

As D'Arienzo (2003) has shown, Emanuele Pessagno (Port. Manuel Pessanha) was Admiral of the Portuguese Sea, a position whose remuneration King Denis (r. 1279–1325) transitioned from payment for services rendered episodically to a system of fealty and salary. The admiralty became a mayorazgo, a hereditary title, that required descendants of Pessagno to swear loyalty to the king and to have twenty Genovese seamen (*sabedores de mar*) ready at all times. In return, the Pessagno family could use an emblem (empresa) as a sign of their house (a ring, a short sword, and the royal arms). However, this mayorazgo only supplemented the family's income. In this way, the "feudal" structure was grafted onto what had been a series of venture capital funds with an ensuing inversion into ownership and management by the Portuguese crown. On the Castilian side, the first admiral of the Iberian Peninsula was Ugo Vento, named by Alfonso X (r. 1252–84) to lead an expedition against the city of Solé in Morocco. Vento was followed by Benedetto Zacaria, also Genovese, who was named Major Admiral of the Sea (Almirante Mayor del Mar) by Sancho IV (r. 1284–95). By the reign of Alfonso XI (r. 1312–50), the position of admiral in the kingdom of Castile had taken on the qualities of the mayorazgo as a hereditary station, as it had with their Portuguese counterparts. For example, Ambrogio Boccanegra, who served Enrique II (r. 1369–79), inherited the admiralty from his father, Egidio Boccanegra, who was Alfonso XI's admiral. As both monarchies employed the venture capital structure to finance their overseas expeditions, the differences between their respective financing schemes remained at the partnership level.

By the mid-fifteenth century, the Portuguese monarchy owned the ships and the Pessagno admirals offered the managerial expertise for sustaining an ongoing naval enterprise. For his know-how and leadership in commercial and military endeavors, the admiral received the carried interest, a fifth part of all booty amassed from infidels and enemy kingdoms, except for slaves and enemy ships, which were claimed by the Portuguese crown after the liquidation of each enterprise. The Pes-

sagnos continued to take freight and sell insurance on their own ships
and transported goods and slaves between Flanders, Lisbon, North Africa, and Genoa on Portuguese navy ships when these were not com-
missioned. Yet the Pessagnos' slave trade, mercenary work, and freight
did not amount to "side businesses" in their position as admirals of
the Portuguese fleet. The admirals' active involvement in the commercial and mercenary networks of continental Europe and North Africa
rather made them all the more valuable to the Portuguese monarchy.
The monarchy also conceded an independent jurisdiction to the admiral's quarter in Lisbon, offering a legal and spatial axis between capital
and sovereignty that Lisbon's inhabitants navigated on a daily basis.[8]
Conversely to Admiral Pessagno's 20 percent share of all navy ventures, the Spanish monarchy received the quinta real, one-fifth of all
booty from conquistas on ships it did not own but whose enterprises it
managed from the Casa de Contratación in Seville, constructed in the
early sixteenth century to manage the exchange of commodities, treasures, and people between Spain and the Americas.

In his contrast between the Portuguese and Spanish modes of conquest, Acosta made no reference to the church's early involvement with
the Genovese venture capitalists before the sixteenth century. Yet the
earliest incursions into the venture capital model on the Iberian Peninsula may be attributed to the church and its ties to the Genovese colony
at Santiago de Compostela. The church provided material and spiritual
incentives to Christians by outfitting commercial and martial expeditions that bore a close resemblance to the usury that its theology
condemned. Offering material and spiritual compensation for sea merchants and men fighting against Muslim populations in the Iberian
Peninsula and northern Africa, bishops were able to procure labor
(among Christians) in order to secure labor and materials (from the
"infidels") for major construction projects. For example, in the twelfth
century, Diego Gelmírez (ca. 1069–1149), bishop of Santiago de Compostela, funded his campaign in northern Africa by purchasing the
ships and paying for a Genovese shipmaster to oversee the military and
trading expeditions. The Holy See supported the voyages by promulgating crusade letters and bulls that allowed the bishops to preach holy
war against the Moors in their dioceses and offer plenary indulgences
to members of the fleet. The bishop received 25 percent of all booty,

in addition to his share as shipowner, which suggests involvement in the fund at both the general and limited partnership levels. However, the sea merchants did not partake of the spoils in labor. All Muslims who had been taken prisoner were handed over to the bishop in order to provide manual labor for the construction of the church dedicated to Santiago.

The church's reckoning with sins, labor, and ships for the construction of, among other structures, a nave at a pilgrimage site brought a measure of sanctity to maritime enterprises that for so long had existed beyond the nomos of the land. After all, the ship as symbol for the church has deep roots in Christological and patristic imagery from the earliest period of Christianity. Not only had the ledger book, as Le-Goff (1988) has contended, influenced the creation of purgatory as the space where service of a spiritual debt involved the activities of both the living and the dead, but the ship of souls tossed on the waves of profanity no longer brought the believers to safe harbor. Instead, the traffic of souls, of believers and nonbelievers, became another currency in the conquests for monetary and spiritual rewards.

Did the Portuguese model, lauded by Acosta, lead to greater political power of the Portuguese monarchs over "their Indies," as the Jesuit scholar suggested? Just as a wholescale comparison between the Iberian empires remains beyond the scope of this book, so too the larger question of the relationship between capital and power may provide the true north but not the final destination of this inquiry. However, the inverse proportions of capital investments do offer a stark contrast in the subjects making claims to "managerial expertise" in Portuguese and Spanish conquests: in Portugal, the quinta do admiral; in Spain, the quinta real; and in Lisbon, the *bairro do admiral,* or admiral's quarter; in Seville, the Casa de Contratación, a royal mini-citadel. During the long sixteenth century, the Portuguese monarch would construct similar mini-citadels, the Casas da Guiné and Mina and Índia to regulate state and private trade in profitable violence. However, the distribution of partnership levels remained largely the same, with the Portuguese monarchs fulfilling their roles as limited partners in terms of their majority stakeholding in any given enterprise.

Taking its cue from the successes of the Genovese model, the Spanish monarchy laid a claim to the quinta and in doing so made a larger claim to "managerial expertise" when overseeing the maritime trading, military, and missionary ventures also known as the Spanish Conquest. The Genovese bankers' network had built its mini-citadels across cities around the Mediterranean and along the Atlantic coasts of Europe and North Africa. It has been widely accepted that though these settlements were profoundly different, their planning reflected the subdivisions of the Genovese republic: *castrum*, *civitas*, and *burgus*. The most strongly defended part of the Genovese *comune* (city) was its commercial core, the castrum, which consisted of a gridded system of urban blocks. This grid extended into the civitas, a second enclosed perimeter that included the buildings occupied by the Genovese urban aristocracy. Beyond the civitas, and always or usually outside the walls, grew the burgus, or *borgo* (town), in a relatively ad hoc manner. In this heterogeneous urban quarter, building construction and daily life were no longer constrained by the grid of the financial center (castrum) but nonetheless revolved around it. In general, the Genovese did not live in the borgo or impose direct rule on the city to which they had appended their fortunes. Juxtaposed to the native city's power center, the admirals' bairros and Genovese quarters employed a strategy of independent management for the commercial and territorial ambitions of their local clients and the comune of Genoa.

In the Spanish mode of conquest, the Genovese bankers' city within a city took on a new aspect in the early sixteenth century with the construction of the Casa de Contratación in Seville, which oversaw the ventures in profitable violence funded and performed by investors of various nationalities in the West Indies. After Seville, the other major banking and seafaring centers were Valencia and Palmas de Gran Canaria. Bankers such as Francesco de Bardi operated in Andalucía, the mid-Atlantic islands, and Santo Domingo and also had personal connections to Christopher Columbus. But the great "revolving door" between enterprise and state could be found in Seville among families of Genovese, Florentine, and Andalusian origins as they jockeyed for positions as financiers and state comptrollers of expeditions. Foreigners and other "undesirables," such as Portuguese conversos, who were nominally prohibited from migrating to the Indies, could nonetheless be called on to invest in venture capital funds. Foreigners were also nominally prohibited from engaging in businesses (*negocios*) and receiving concessions (*concesiones*) and shareholding contracts (*asientos*). However, procurement of a "naturalization" certificate allowed bankers of different nationalities to make loans and occupy bureaucratic posts as a way for the crown to service its debt (Sanz Ayán 2004, 37). Thus the foreigner prohibition and the naturalization exemption serve as another example of making recourse to the exception when reckoning with limited access to capital and subsuming that capital in the service of empire.[9]

The venture capital model serves to elucidate the Spanish sovereigns' contributions to the imperial enterprise and to highlight some glaring conflicts of interest. For example, it has been noted that before the large-scale extraction of silver from Potosí, expeditions to the Spanish Americas operated at a net loss (Fisher 1977, 22–23). However, in venture capital funds it is possible for the enterprise to go bankrupt but for the general partners and some limited partners to receive carried interest. Within these parameters, the crown and the church would be the general partners aiming to secure rapid scalability but also sustainability from the concatenated investment in a series of funds (conquistas and descubrimientos). As providers of managerial expertise, they did not make the major capital investments but organized and distributed the enterprises through tangible and intangible forms aimed at rapid expansion (in the face of competitors) and sustainability. Yet sustainability and scalability compete for resources in any enterprise and come into conflict among the priorities pursued by individuals within the venture capital hierarchy. As bankers and entrepreneurs organized transatlantic trading companies in Seville, the crown responded by creating its own mini-citadel that structured the commercial activity of the financiers (*cambiadores* or *hombres de negocio*) and merchants (*mercaderes*). Investors made contributions to the funds both in coin and in kind—such as grains, animals, cloth, and weapons—leading to uncertainty in determining the relative values of commodities, specie, and, thus, distribution of ownership in any one enterprise. This uncertainty stretched across enterprises and into the fiscal operations of taxes and tribute. A *banquero* (banker), mercader, or hombre de negocio could expect the award of an asiento as both loan guarantee and debt service, a glaring conflict of interest. The award of an asiento to tariff or tribute collection in specie or in kind gave the *asentista* the right to collect and enforce collection in the name of the state and, in turn, to receive repayment for an outstanding loan to the crown from the moneys or tribute collected by the asentista.[10] The conflicts of interest inherent in such a system of repayments were apparent to all parties involved.

At the limited partner level, cambiadores and mercaderes of Seville lent their expertise to the crown: refining gold destined for the House of Coin.[11] The term "banquero" first appears in Seville in the documents outlining the liquidation of a fund owned by two Genovese brothers, Batista and Gaspar Centurione. In their founding partnership document (*convenio*), they described themselves and their company as a trading and banking company (*compañía en el banco y cambio, y fuera de él en cualquier manera*) or a currency exchange owned by the bankers Batista and Gaspar Centurione (*cambio de Batista y Gaspar Centurione, banqueros*) (Otte 1978, 93). Though this partnership (societas) lasted for only three years (1507–11), Gaspar Centurione formed another short-lived partnership with Juan Francisco Grimaldi (1511–14). In this way, the activities of the Centurione and Grimaldi families in Seville reflect the classic pattern of venture capital funds: a rapid succession of three-year ventures that create several, accelerated temporal horizons for profit within a decade. Yet risks for the bankers were plentiful as well. For example, the failure of one expedition associated with the fund owned by Díaz de Alfaro, Rodrigo Iñiguez, and Bernardo Grimaldi, brother of Juan Francisco, left Grimaldi bankrupt in 1510.[12] Until that time, Bernardo Grimaldi had been the most active of all Genovese merchants in Seville.[13] In the face of the rapid and scalable expansion of funds profiting from violence overseas, it was the task of the Casa de Contratación, the Council of the Indies, and the church to ensure continuity across the various ventures, to thread a grand narrative of empire from the episodic expeditions: from a series of conquistas, the larger narrative and legalistic arc of the Spanish Conquest.

The appreciation in value of science and technology took on an institutional form with the creation of such positions as the *piloto mayor* (first held by Amerigo Vespucci) and the Sailing Academy (Universidad de Mareantes). As the state reduced capital investments in each enterprise funded by various banqueros, it built an apparatus that sought to reduce inefficiencies such as the loss of cargo or life due to the lack of expertise of navigators or the temerity of expedition leaders. The Universidad de Mareantes provided instruction and certification in the use of instruments, such as the astrolabe, and cartography, creating knowledge for best sailing practices (Cervera Pery 1997, 91–128).

On the one hand, bureaucracy flourished in an attempt to reduce the risks of failure or corruption in each imperial enterprise, as each item in every contract was signed by the general and limited partners. On the other hand, bureaucracy acted in the service of each enterprise's stakeholders, many of whom also occupied official posts. It was not unheard of for the state treasury to act as a guarantor of loans made by private bankers to the Castilian crown.[14] Had they not incurred material risks, however, these enterprises would have been in danger of committing usury.[15] Yet if the crown protested, perhaps too much, that its share of material risk must be reduced, why then does the crown's "managerial expertise" carry such a heavy price? Why do some types of risk underwrite the moral propriety of economic activity while others reflect poorly on the moral health of diverse subjects? After all, risk taking was not without its moral faults. Courting risk in games of fortune, such as cards or dice, was frowned on and, as in Ferdinand II of Aragon's contracts with explorers, strictly prohibited.[16] As in the 1542 laws concerning the encomiendas and slavery, risk was not defined per se but functioned as common currency in the moral economy of empire. Like many coins in circulation at the time, ensuring the equivalence between the face value and the intrinsic value of *peril* was an activity fraught with anxiety.[17]

ACCOUNTING FOR PROFITABLE VIOLENCE

Conquest underwent lexical renewal and demise in less than a century. This transformation mirrored the accelerated temporal horizons of venture capital funds. As Gibson (1977, 20) contends, the entry of *conquista* into the early modern Castilian lexicon reflected its Latin etymology: the past participle of *conquaestare*, *conquista* signified the result of an exploration or discovery, often in lands outside the Iberian Peninsula. By 1611, Sebastián de Covarrubias y Orozco (1539–1613) would make explicit the violence of discovery and exploration in lands inhabited and ruled by others. In his *Tesoro de la lengua castellana o española* (1611), Covarrubias defined *conquista* as "pretender por armas algún Reyno, o estado" (to feign or expect [to achieve] by force of arms a kingdom, or state). The verb *pretender* reinforces the metalepsis of foundational violence "by force of arms"; without the violent reinforcement, the pretensions—fictions but fictions that entail expectations—would remain unrealized. Similarly, the lexicographer alludes to "kingdoms" or "states" as the result of conquista's violent enactment. The definition thus glosses over the disruption of orderly, perhaps even civilized societies implicated in conquista's pretensions. The figurative leap of the verb *pretender* suppresses the transition from violent possession (merum imperium) to the constitution of a kingdom or state. In Covarrubias's muted allusions to foundational violence (i.e., *pretender por armas*) can be heard the faint echoes of Bartolomé de Las Casas's diatribes against the "dichas conquistas" in his *Brevísima relación de la destruición de las Indias* (1552). When did *conquista* become "conquista"? In other words, when did the utterance of *conquista* become a self-reflexive exercise, where the speaker felt the need to justify his or her use of the term?

Conquista's fortunes, in moral and material terms, were tied to the particular structure of venture capital funds in the invasion of the Americas. The word *conquista* may have fallen out of favor by the end of the sixteenth century, but its mode of operations left a legacy of discourse and narrative that continues to haunt modernity. As many other commentators on the *Ordenanzas* have observed before me, despite all the rhetoric of novelty and reform, these laws did not change the methods of empire, *grosso modo*, at the close of the sixteenth century. Why, then, did *conquista* continue to be so grating a term that, for some, was nevertheless so inspiring? Popular wisdom, as recorded in the entry *desorden* (unruliness) in the *Diccionario de Autoridades* (1732), may provide a clue to *conquista*'s continued salience for the ordering compulsion of the empire: "el desorden trae el orden. Refrán con que se da a entender que los gastos supérfluos y prodigalidad acarréan pobréza y miséria: y ella obliga à la moderación y buen gobierno" (unruliness brings order—a saying that suggests that superfluous expenses and wastefulness entail poverty and misery, and this compels moderation and good government). It would seem that order and good government would be unimaginable without its antithesis: unruliness and excess. And yet Las Casas, and even the Spanish crown by 1573, had come to associate conquest with its supposed opposite, frenzied violence. The blurring of formerly antithetical values—conquest and frenzy, order and unruliness—was uncanny and a source of great anxiety.

At the same time, *empresa*, the word commonly associated with "business" or "enterprise" in contemporary Spanish, suffered a transformation similar to that of *conquista*. *Empresa* had been widely understood as an "activity with a purpose" or "activity to an end"; knights of romance narratives made empresas, but so did day laborers (Vilar 1978, 243). By the close of the sixteenth century, acceptance of the word became increasingly circumscribed, limited to the sphere of commerce. Yet commerce and violence, as Vilar (1978, 245) suggests, could be embodied in the same figure: "Cristóbal Colón, último gran empresario caballeresco, primer gran empresario al servicio del capital" (Christopher Columbus, last great knight empresario, first great empresario in the service of capital). How did *empresa* and *conquista* dovetail and then part ways? It seems almost too easy to signal Columbus as the beginning and the end of eras in capital and chivalry. Indeed, the definition of *empresa* by Sebastián de Covarrubias underscores its emblematic function, created to fulfill a particular end:

> Emprender, determinarse a tratar algun negocio arduo y dificultoso, del verbo Latino apprehendere, porque se le pone aquel intento en la cabeça, y procura executarlo. Y de alli se dixo Empresa, el tal cometimiento: y por que los caualleros andantes acostumbrauan pintar en sus escudos, recamar en sus sobreuestes, estos designios y sus particulares intentos se llamaron empresas: y también los Capitanes en sus estandartes quando yvan a alguna conquista. De manera, que Empresa es cierto símbolo o figura enigmatica hecha con particular fin, endereçada a conseguir lo que se va a pretender y conquistar, o mostrar su valor y animo. (1611, 345)

> [To undertake (*emprender*), to resolve to do an arduous or difficult *negocio*, from the Latin *apprehendere*, because once the intention is placed in one's head, [one] intends to execute it. From this it was said *Empresa*, this undertaking: and since the knights errant would paint their shields, embroider their clothes [with it], these designs and particular plans were called *empresas*; and also the Captains [used them] in their standard when they went on *conquista*. In this

way, *Empresa* is a certain symbol or enigmatic figure made to a particular end, raised in order to achieve what will be feigned and conquered, or to show valor and intent.]

Empresa, or enterprise, as a fait accompli corresponds to events that have ended, that are narrated in the preterit tense, much like the narratives of the exploits of the knights errant (*caballeros andantes*). However, in the lexicon, the time of *conquistar* and *pretender* remains open-ended. *Conquistar*, *pretender*, and *emprender* are used synonymously; *conquista*, *pretensión*, and *empresa* are thus corollary figures and fictions of one another. In this way, both the entrepreneurial and the emblematic functions of *conquista* remain viable for Covarrubias at the turn of the seventeenth century. More than a century after the *Tesoro de la lengua* was published in Madrid, the academics of the Real Academia included the business venture as a second entry, as an extension of the first acceptance of *empresa*, as emblem or sign.

> La acción y determinación de emprender algun negocio arduo y considerable, y el esfuerzo, valor y acometimiento con que se procura lograr el intento. (*Diccionario de Autoridades*)
>
> ———
>
> [The action and decision to set forth (or undertake) an arduous *negocio* worthy of consideration; and the effort, valor, and undertaking with which the intent is procured to be achieved.]

Enterprise thus acts as a hinge between a negocio, or business—busy-ness, the negation of *otium*—and a sign used to signify the goal of an "arduous," "difficult" negocio in the process of becoming. As in the case of conquista, an action's intention and an action's result are conflated. Yet empresa also invokes the emblem, the "self-fashioning" of Renaissance subjects within the parameters of socially acceptable standards. At the same time, empresa, an action in the service of a prize, conflates standard with standard-bearer. As such, as an enterprise that required a degree of self-reflection, it emphasized the subject's identity and action in juxtaposition to the empresas of others. The self-fashioning involved in empresa offered the possibility of rupture but also continuity with the past. Thus, the character of empresa and its agents involved the imagery of a battlefield and combative interests.

Ideally, however, conflicting accounts of different empresas could be forced to agree.

Accounting for enterprise became an enterprise in and of itself. Entrepreneurial genres of introspection involved double-entry book-keeping but also narrative accounts of verbal exchanges with other merchants, as well as longer, multiyear narratives. What Baldassar Castiglione (1478–1529) was to the self-made courtier, Benedetto Cotrugli (d. 1468) was to the merchant, offering these words of advice to the would-be "perfect" merchant in his book *Della mercatura et del mercante perfetto* (1573): "When you see a merchant to whom the pen is a burden, you may say that he is not a merchant" (quoted in López and Raymond 2001, 375).[18] Cotrugli's insistence on self-inspection in different forms of writing—the *quaderno*, *giornale*, and *memoriale*—with various temporal horizons (daily, monthly, and yearly) offered episodic opportunities to juxtapose and reconcile contradictions. The state of a merchant's *credit*, that is, the belief (from *credo*) of his peers in his ability to keep his word, in turn, increased or alternatively decreased access to capital investment.[19] Cotrugli's "perfect merchant" was a humanist, lover of the arts and writing, knowledgeable in local customs and laws, and a master at defending his own interests while recognizing the rights of others even when transactions involved a zero-sum game. Thus the perfect merchant shows a certain effortlessness, akin to *sprezzatura*, in masking conflicts of interest. In effect, Cotrugli's manual of perfect (self-)merchandising anticipates the concerns and machinations outlined by Castiglione's famous *Cortegiano* (1528), though the correspondences made between word, honor, value, and comportment are unequivocal in Cotrugli's book. And yet narrative contradictions do arise when attempting to reconcile the discrepancies between daily log and historical account.

The merchant's faith in his ability to reconcile accounts was challenged in the New World, as the need to document went hand in hand with the prerogative to erase. Such is the case of Christopher Columbus and his *Diario a bordo*, as edited by Las Casas, and the latter's *Historia de las Indias*, neither of which was published in the sixteenth century (see ch. 3). However, the first narrative to emerge from the New World was Columbus's *Carta a Luis de Santangel* (1493), whose translation into Latin was widely published and retranslated throughout Europe. The *Carta* launched, as Stephen Greenblatt (1991, 81) has argued, the conjunction of the most "resonant" legal ritual, that of possession, with the most "resonant" emotion, that of marvel. Letters written by investors and entrepreneurs in the West Indies were modeled on Columbus's first letter, intended to generate *credito* from that foundational conjunction between marvel and possession.

Successful practitioners of the commenda, societas, or triple contract in the West Indies could commandeer 100 to 200 percent profits per fund, according to these letters. Speaking of a Florentine merchant who claimed to be receiving such returns on his investments in the Americas as early as 1502, Piero Rondinelli urged his fellow countrymen to invest in Bardi's new funds because "Francisco de' Bardi s'à a fare riccho a meravaglia" (Francesco de' Bardi knows how to get rich marvelously) (quoted in D'Arienzo 2003, 227). Rondinelli's admiration for Bardi's know-how adds another element to the image, in the process of becoming, of America as cornucopia: it is not only a place where commodities and labor abound, but also where capital flourishes. This reconciliation of the traditional oxymoron of usury—unnatural usufruct—uncovers the vein of promised returns on investment that is the subtext of Columbus's first letter from his first voyage. In his first letter to Luis de Santangel (d. 1498), Ferdinand II of Aragon's finance minister and the main sponsor of the admiral's first voyage, Columbus modeled his experience of marvel for readers, based on his account of his reactions to the natural bounty of the two main islands he had been observing:

La Española es maravilla: las sierras y las montañas y las vegas y las campiñas y las tierras tan hermosas y gruesas para plantar y sembrar, para criar ganados de todas suertes, para edificios de villas y lugares. Los puertos del mar, aquí no habría creencia sin vista, y de los ríos muchos y grandes y buenas aguas, los más de los cuales traen oro. (1984, 141)

[Hispaniola is a marvel: the hills and forests and meadows and the countryside and the lands that are so beautiful and loamy for planting and sowing, for raising all kinds of cattle, for building villages and places. The ports, which must be seen to be believed, and the many rivers, wide and deep, many of which carry gold.]

Columbus's letters and *Diario* identify the opportunities for trade, mining, agriculture, and enslavement as marvels to behold. Yet Columbus was more cautious than Rondinelli would later be in his appraisal of opportunities for making money from money. Soon, however, the oxymoron of flourishing capital gave way to a rapid reconciliation of contradictory concepts, and the original, more cautious approach to usury by Columbus is superseded by comparisons between natural and unnatural economic activities, a metalepsis without qualms.

Not only might one marvel at the fruits of the earth in the Americas; one might also marvel at the knowledge of a certain class of men, like Piero Rondinelli, who knew how to make money — no longer sterile — fruitful. The cornucopia of capital was not natural to the Americas; capital was marvelous in the hands of a few who, in the short term, controlled the capital flows into and out of the continent. The metaphors of the land's bounty and the rivers' depths were applied to that most imperial of rhetorical figures: the *translation*, seemingly without contradiction. The dearth of coin circulating in the colonies and uncertainty about the value of commodities in exchanges within the American colonies added to the circumscription of America as a place where money went to multiply and leave. The dearth of coin can be contrasted to the preponderance of accounts: *diarios* (diaries), ledgers, ship manifests, *crónicas* (chronicles), and capitulaciones, or contracts, that all aspired to a settlement — in specie, in kind, and in tribute — at the close of each fund. Losses and gains became commensurate items in the various methods for documenting profitable violence. Tabulated risks could evolve into narratives of great riches or increased material liabilities that, perhaps, were offset by the moral gains in the behavior of the subject.

ACCOUNTING FOR RISK

Risk taking and stakeholding in an enterprise were not only a means for making a profit but also, and equally important, a foil against committing the mortal sin of usury. During the second half of the sixteenth century, financiers such as the Fuggers of Aubsburg, who financed the Hapsburgs' wars on the European continent, sought greater clarity in canon law with regard to loans that charged interest as insurance, a setup known as the triple contract (Noonan 1957, 206–33). Indeed, the risks of loss of life and property during sea voyages also made Scholastic thought more amenable to the distinction between use and ownership when such risks were pooled in the societas. In order to have a societas—a term that continues to have currency in the Romance languages, for example, *sociedad* in Spanish—two or more persons form a union of their capital and/or skills for a common purpose.[20] In the fifth of the *Siete Partidas*, Alfonso X commends the societas or compañía as a union of two or more men who seek to profit together in a partnership intended to provide great benefits to all partners. However, partnerships for usury (*dar logro*) were prohibited, although the law does not go into greater detail regarding the difference between *dar logro*, "to generate interest," and *recibir provecho*, "to profit" (Partida 5, Título 10, Leyes 1 and 2).

On the one hand, as Noonan (1957, 133–53) has shown in *The Scholastic Analysis of Usury*, the equitable distribution of risk was the main factor used by Scholastic thinkers to distinguish between a usurious loan and a legal partnership, yet this appreciation of risk contradicted Aquinas's axiomatic definitions of property that did not distinguish between the use and value of property. On the other hand, the moral value of risk taking led Scholastic thinkers, even Aquinas, to contradict their own definitions of property and propriety, which they argued in terms of use and usufruct. Though Aquinas made no distinction between the *use* and the *ownership* of money, it is precisely on the basis of such a distinction that he defended the societas, even though the capitalist relinquishes *use* of his money to his partner but not ownership thereof in a partnership. Thus Aquinas observes in the *Summa Theologiae*, "He who commits his money to a merchant or craftsman by means of some kind of partnership does not transfer the ownership of his money to him but it remains his; so that at his risk the merchant trades, or the craftsman works, with it; and therefore he can licitly seek part of the profit thence coming as from his own property" (2.2.9.78.2, obj. 5). The introduction of a third element, *peril* (and its price), drives a wedge through the equivalence between use and ownership and so allows Aquinas to accept the societas, seemingly without contradiction.

Aquinas even availed himself of risk for a definition of ownership to which his earlier, axiomatic definition based on the understanding of use as ownership does not. Thus a partnership in which one partner puts up capital and another labor ("sweat equity," as it is known today) is not usurious if the risk is shared equitably, even though such a formulation would have been designated usurious in land-bound contexts (Noonan 1957, 143–45).[21] This breach in the continuum between use and ownership went to the heart of the prohibitions against usury, especially in seafaring enterprises. When the exception to usury was coupled with the other exceptions to jus gentium, such as freedom of movement to trade and proselytize, whose agents' efforts could, in turn, enjoy a military "defense," the militarized trading company enjoyed a moral prerogative that became rapidly scalable, globally.

In Acosta's lifetime, the last major overhaul of the Spanish laws governing the Indies were the *Ordenanzas* of 1573, which, as we have seen, notoriously outlawed the use of the word *conquista* when referring to the economic and military ventures that could also be described as *descubrimientos*. However, the *Ordenanzas* did not restructure the Spanish crown's investments in specie and in kind. Rather the laws kept the crown's initial investment at a minimum, maintaining a common practice that had been in effect since 1504 when the Castilian crown officially abolished "monarchic capitalism" in the Portuguese vein.[22] Each capitulación, or contract, signed between the crown and the crew in the Casa de Contratación outlined subordinate ventures of the main enterprise. Line by line, the identities of Indigenous persons, as new subject or slave, were defined in lockstep with the itinerary that delineated the geographic area and its peoples subject to the conquest at hand. Provisos inveighed against weighing down the ships with too much cargo, set salaries for clergy, fixed the different distributions of booty at sea and on land, established the lease terms of encomiendas, and prescribed grains to be cultivated overseas. These *items* were not only orders that expected compliance from subjects who were also business partners; they also served as indexes for the underlying stakeholding apparatus, not only in each individual fund, but also in the imperial enterprise as a whole.

INTEREST AND *INTER ESSE*

Venture capital offers great rewards at great risk to its investors for it traffics in a paradox of great import to the "Conquest of the Indies":

ownership in something as yet to be discovered, invented, or created. Unlike other business ventures, the stakes are drawn in an enterprise before it begins or is even completely understood. Failure of the enterprise brings great losses, but success brings commensurate gains, often in novel ways. The promise of the novel enterprise—and to own a piece of "it" before *it* comes to fruition—draws investors to its cause. As such, this type of business venture arises from a consciousness not readily explained by Scholasticism's powers of the soul (memory, understanding, will). It also calls into question the argument made by Edmundo O'Gorman in *La invención de América* (2006) that unless you *intend* to discover an entity, it has not been discovered by you. Much like Juan Huarte de San Juan's revision of Scholasticism's powers of the soul, which replaced *la voluntad* (the will) with *la imaginativa* (the imagination) during the second half of the sixteenth century, venture capital privileges the imagination as the faculty for appropriation. Hence, stakes are drawn in an enterprise before it begins or is even completely understood. The "process" of exploration, as the cultural geographer John Allen (2003, 56) argued, "is conditioned by the imagination" and the interplay between received and empirical knowledge of *terrae incognitae*. Like exploration, venture capital is an intersubjective process of creation and appropriation. As discussed above, venture capital's increasing acceptance as a legitimate means for profit drove a wedge between use-value and ownership in canonical definitions of property and moral propriety, an altogether unorthodox distinction that relied on, among other things, the subject's penchant for taking risks.

Who were the venture capitalists? In fifteenth- and sixteenth-century Spain the limited partners were wine, grain, or wool merchants who had the international networks and disposable income to invest large amounts of capital or goods in short-term conquista funds. In addition to being merchants or merchant capitalists, as Braudel described them, they often held and pursued asientos and *juros* (annuities) in city governments or in the Casa de Contratación.[23] As creditors of the crown and tax and tariff collectors, this often led to conflicts of interest. Yet these limited partners were only one piece of the venture capital puzzle. However, despite the corporate nature of venture capital and the conflicting interests at play in each enterprise, narratives of empresas portray the entrepreneur as a solitary figure who follows a calling.

A vocation limited to a reduced number of people within the venture capital fund as a whole, entrepreneurship dominates narratives of both conquest and venture capital. Vocation, from *vocare*, "to call, summon," implicates both the act of calling and the state of being called, an intersubjective dynamic that bears some similarities to the duality of love, lover and beloved. Yet proponents of capitalist vocation, such as Max Weber and Schumpeter, have turned the tables on the Aristotelian binary that favors the "active nature" of the lover over the passive nature of the beloved.

What are the implications of visualizing entrepreneurs as passive agents, an oxymoron of sorts, that is, as *being called*? This calling, as Weber argued in *The Protestant Ethic and the Rise of Capitalism* (2009b), has a particular voice: an intersubjective drive for material and spiritual redemption that manifests capitalism as both choice and thrall.[24] Invocation confuses the "calling" with "being called," yet narratives of entrepreneurial efforts offer insight into the upheaval of the active/passive binary in capitalist consciousness. Navigating the dire straits of transatlantic commerce, empresarios heeded the siren's call, cupiditas incarnate, then courted the risk, reaped benefits, and lived to tell the tale. The rise of romance and conquest narrative during Europeans' quests within and beyond the bounds of Hercules's columns from the twelfth century to the modern period, as explored by Nerlich (1987) and Campbell (1999), represents the subjectivity of booty-driven capitalism.[25] However, even forms of "primitive" or "adventure capitalism" were corporate entities made up of many individuals that responded to *calling(s)* that incorporated conflicts of interest and varying levels of liability among their members.

Here, the etymology of *interest*, from the Latin *inter esse*, or interbeing, refracts onto the processes of incorporation in the construction of companies and empires. The legal personhood of corporate entities masks their ambivalence, or the coexistence of at least two opposed and conflicting wills. As Weber concluded in his *History of Commercial Partnerships in the Middle Ages* (2003), the law struggled to define the juridical personhood of these partnerships in the context of international commerce. As usual, the law was one or several steps behind commercial practice. Thus the *Statutes of Genoa* (1588–89) codified into law the multifaceted hierarchy of venture capital firms more than a century after the household and the physical warehouse (*bottega*) had given way to the various accounts in capital held by each firm. The terms of liability changed, and multiple personae emerged. Partners of a societas were only liable for those contracts signed by another partner with the power to represent the other members of the firm. The *duplex persona* (double personhood) emerges to account for distinctions between *propia negotia* (personal business) or *quorum nomina expenduntur* (those whose names have been left out). In this way, the identities and "personhoods" are contingent on their identification and incorporation into each fund via the contract. And yet the metaleptic trope for personhood in the societas, which employs synecdoche, unless made explicit by the individual contract, both loses its function for representing enterprises in liability cases and has an exaggerated role in the figures of entrepreneurs as heroic, solitary figures. A totality, or corpus, is affirmed in the legal fictions of the societas, but its personhood remains in constant flux. This explains, in part, the incongruity of the many figures of metalepsis in representations of venture capital enterprises.

The rift in representation between the firms that employ the personhood of the *corpus mysticum ex pluribus nominibus conflatum* (a juristic person comprising many names) and the narrative of their agents' actions widens in adventure-writing ideology. As the law attempts to keep abreast of liable persons and personhoods in international commerce, the multifaceted duplex persona sets out on heroic, individualized quests. Rather than tell of the exploits of hybrid corporate beings, narratives of empresas represent knights fighting the monsters without, though evident sutures in heroes' characterizations warn of the monster within. One such suture is the topos of the double name or translated names of the hero in adventure writing, a narrative convention used in Gonzalo Fernández de Oviedo's *Claribalte* and explored at length in chapter 3.

The possibility of Leviathans venturing out on quests seems almost nonsensical, or topsy-turvy.[26] In contrast to the bricolage characteristic of mythological thought, as elaborated by Claude Lévi-Strauss (1962), the duplex persona of the empresa, though monstrous, is cut from the same cloth of goal-oriented and capital-driven adventures. In this sense, the uniform delineations of corporate personhood in venture capital funds are closer to Miller's (1977–78) concept of the "suture" to trace the subject's insertion, via signifier, into the symbolic order of language. That the monstrous nature of venture capital funds should split into inordinate desires, on the one hand, and the ordering impulse, on the other, speaks to the conflict embodied by corporate beings that, nevertheless, attempt to erase their inter esse while pursuing their interests.

It is no wonder then that the most ardent critic of the system of martial commerce in the West Indies, Bartolomé de Las Casas, would withhold the names of the individuals whose actions he condemns in the *Brevísima relación de la destruición de las Indias* (1552). To do otherwise would succumb to the logic of the commenda and its system for limiting liability for the general partners, who, at the apex of the venture capital firm, kept separate accounts for each fund (i.e., the limited partnerships). This hierarchy of partnerships—the distinction between the general and limited partnerships—not only limited their liability for each failed fund but also allowed them to profit from each and every success. The multifaceted agency of general and limited partnerships in imperial enterprise makes it difficult to define the moral personae of these corporations in order to adjudicate liabilities. Thus Weber's (2009a, 78) definition of the modern state as "the monopoly over the legitimate use of physical force within a given territory" falls short of describing the interaction between capital and violence in "foreign" conquests. In this sense, discussions of corporate, legally sanctioned violence, such as conquista, that fail to engage with piracy, mercenaries, and banditry remain circumscribed by the tautology of legally sanctioned violence. The shift in discourse to (en)force a reconciliation between antithetical love and interest falls squarely within this tradition.

THE MORAL PRICE AND MATERIAL PREROGATIVE OF PERIL

For the church, the crown's other general partner in imperial enterprise, liability, profit, and peril were values traded in a moral and material economy, a closed system whose political theology reproduced on a global scale during the long sixteenth century. Paradoxically, as Anghie

(1996) has argued, the respect for local values, customs, and laws—known as jus gentium—served to propagate Christian, European beliefs as basic tenets shared by all peoples. As a legal term, "jus gentium" originated in Roman jurisprudence and refers to common law or local customs that may be recognized by imperial magistrates as law as long as they did not conflict with imperial law. However, the further development of jus gentium by jurists of the Dominican order of the School of Salamanca, especially Francisco de Vitoria, as the source for the modern concept of national sovereignty cannot be understood outside of the Spanish colonial project overseas. Namely, the invocation of rights through "natural law" imposed a standard of conduct from a Christian perspective that guaranteed universal rights, including the right to free movement across lands of different peoples and cultures for the purposes of trade and evangelization.

Anghie (2007) has also traced the genealogy of the "war on terror" and free trade agreements to the original Spanish colonial paradigm that arose out of the School of Salamanca, especially Vitoria's lectures, given in the early to mid-sixteenth century. Seen in this way, Christian-inflected jus gentium, at the roots of international law today, emphasized the "universal" attributes of "all" peoples, such as their "natural" desire and right to trade their wares (*jus comercii*) and to move "freely" for this purpose (*liberum commercium*). Coupled with the Christian prerogative to move freely for the predication of the "true" faith, the School of Salamanca's recognition (or, rather, creation) of a "universal" desire to trade, when couched in the terms of imperial respect for local mores, served as a moral buttress to practitioners of profitable violence in the Indies. Though Vitoria and others would question the "just war" rationale proffered by, among others, Juan Ginés de Sepúlveda, their articulation of jus gentium and certain universals—notably, the desire to move freely in order to trade and proselytize supposedly shared by all peoples—armed missionaries and merchants with the rhetoric of violence performed in self-defense.

The free movement exception drove the wedge behind which Christian empire could impose itself on local sovereignty, globally. In light of this exception, the polemics surrounding *de jure* possession of the Indies—whether or not the Papal Donation was valid, if one could wage just war against the Indigenous peoples for their barbarous customs—directed the surface of juridical discourse in Salamanca and Alcalá de Henares in the sixteenth century but skirted the price of peril, to life and to possessions, and what the church and crown, at the general partnership level, were willing to pay *morally* given the generalized assumption that de facto empire—the material use of force—could accompany efforts to trade and evangelize. At the limited partner level, the conquistadors could justify their violence in pursuit of profit, and their creditors' profits could in turn be justified by the threat of losses to European lives and property in overseas ventures. But a paradox emerges from this tautology surrounding peril, one that fuels the anxiety around the "so-called conquests" both at the general and the limited partnership levels. Incurring peril to life, limb, and property in business partnerships may have allowed entrepreneurs and investors to largely avoid charges of usury; however, the managerial expertise afforded by the crown and the church also aimed to keep these risks to a minimum. If, on the one hand, the status of native peoples of the Americas as potential new subjects and neophytes gave this ongoing partnership a shared purpose, on the other, the coupling of missionary work with violence against future Christian brethren raised another specter of peril: the souls of all Christians involved in the imperial enterprise. The labor-religious tutelage exchange, known as encomienda, made this peril highly salient because their mismanagement in the Spanish Indies had served to highlight its structural contradiction: former conquistadors received compensation for their past performance based on leases of Indigenous labor and tribute awarded by the Spanish crown in order to steward Indigenous peoples in Christianity, even though they were not charged with the actual missionary work and were soon required by law, after 1526, to keep their distance from their wards. By the close of the sixteenth century, an astute observer of the situation, like José de Acosta, could recognize some of the structural contradictions behind the Conquest, not least the negligible investments made by the Spanish crown in each venture, which he had interpreted as lesser political authority by the imperial sovereign. Yet Acosta's insights into the causes of the encomenderos' rebellions omitted the rapid scalability afforded by the venture capital system of Spanish conquest. In this sense, symbolic and administrative power wielded by the general partners in conquest, the crown and the church, perpetuated the cycle of debt and violence around the tropes of peril and protection.

This rhetoric of defensive action to "protect" missionaries and merchants from the perils of preaching or trading within the larger Christian imperial project is analyzed in greater detail in chapters 2 and 3. For now, I would simply note how "the price of peril" shifted in its moral valence during the sixteenth century. At first, the peril to life and livelihood of entrepreneurs and partners who had invested in overseas ventures justified the material and capital gains that would have been labeled usurious in land-bound contexts. At the same time, peril to life and limb by the very same subjects justified the use of violence to "defend" their ventures overseas. It also placed the church in the unique position of both arbiter and partner in moral and material risk-taking activities that were performed, quite literally, "on the same boat."

The church as vessel for these souls no longer sought safe harbor. It rather projected itself as underwriter for the exchange of goods, labor, and indulgences in which it had much to gain in material and spiritual terms from believers and nonbelievers alike. In effect, the church had become party to its own "triple contract." The spiritual insurance provided by the church would nonetheless, paradoxically, imperil its own powers in matters of the life and death of the soul when it sought to adjudicate as well as participate in the Conquest of the Indies. Moreover, accounting for credits and debits between infidel labor and believers' indulgences became infinitely more complex when the discourse of the enterprise shifted to incorporate Indigenous neophytes, and everyone, including the former enemy, was said to gain. The contradictions that emerge in the contracts and laws drawn up to reckon with new Indigenous subjects, slaves, and neophytes in the Americas is discussed in the next chapter.

CHAPTER TWO

Contracting Love Interests

Undoubtedly peace is the universal good for all things and, thus,
loved everywhere by all who see her. And not only by her[self], but
even by the sight of her image she enamors them and arouses the
desire to resemble her, for all things favor her easily and sweetly. . . .
Because the merchant sails the seas in order to make peace with his
desire, which both compels and makes war with him.
—Luis de León, *De los nombres de Cristo*

The conquistadors and their general partners sought out indios, either
to subject or enslave. Their projections for profitable violence did not
envision deserted landscapes as areas fit for conquista unless they could
be settled, preferably with the labor of Indian men, women, and children. How do we account for the difference between the merchant,
whose wanderlust saw support in the "free movement for trade" doctrine, and the nomad, viewed as a vagabond who must be corralled? The
merchant follows paths that are subordinate to destinations, whereas

the nomad has a territory; he follows customary paths; he goes
from one point to another; he is not ignorant of points (water
points, dwelling points, assembly points, etc.). But the question is
what in nomad life is a principle and what is only a consequence.
To begin with, although the points determine paths, they are
strictly subordinated to the paths they determine, the reverse happens with the sedentary. The water point is reached only in order

to be left behind; every point is a relay and exists only as a relay. A path is always between two points, but the in-between has taken on all the consistency and enjoys both an autonomy and a direction of its own. The life of the nomad is the intermezzo. (Deleuze and Guattari 1987, 380)

As late as 1573, the nomadic relay continued to be seen as an impediment to the progressive spread of Spanish Christian civilization and the *vida con policía* (civilization). The Spanish law's remedies are illuminating in this regard, as they seek to replace the relays of the *indio vaco* (vagabond Indian) with the endpoints of the *indio vaquero* (cattle-herding Indian). For example, in an apropos amphibology, the law required that nomads both corral cattle and be corralled by their colonizers: the indios vacos were to herd the cattle within an encomienda for two lives, or three lives if the settlements were for "cabeçeras" or "puertos para nos," that is, for the crown in a *repartimiento* (Morales Padrón 2008, 58). The enjoined endpoint—*indios vacos vacando* (cattle-herding vagabond Indians)—produces an uncanny conceit in Spanish, familiar insofar as it is strange, to borrow from Freud's classic essay, in that it collapses and reproduces the stereotypes about Spanish versus British, and later "American," modes of conquest. Whereas the Spanish sought labor and neophytes, the British, and later Americans, sought land. As argued by Saldaña (2016, 1–31), the stereotypes trafficked along the Mexico-U.S. border about the neighboring nation-states—one teeming with Indians, the other deserted—are belied by their shared history, but, perhaps more important, they underscore the dependence of dominant national identities—and their concomitant narratives—on the presence/absence of the Indigenous other.

Beyond the troubled and troubling border along the Rio Grande, other places also point to dependence on the indigeneity of others by various competing empires. With the exemplum of the Bermudas in *The Singing of the New World*, Tomlinson has contended that uninhabited spaces were largely avoided by Spanish, English, and French explorers throughout the sixteenth century, unless they were lured there by natural resources. However, even a strategically situated archipelago in the Gulf Stream could be largely "shared" by Europeans in transit, by ships laying anchor to replenish freshwater and other supplies as long

as there were other places, with peoples, to be invaded; the lack of Indigenous presence made that land strangely undesirable to new claims
by imperial entrepreneurs, at least in the beginning (Tomlinson 2007,
1–8). Or to repurpose the terminology of Deleuze and Guattari, the
nomadic intermezzo could serve imperial enterprise at certain points.
Whether the dominant narrative stressed native resettlement or exter-
mination, the imperial enterprise was premised on what the Indigenous
others' future should look like but always in relation to their invaders.
That Indigenous peoples would give purpose to these projects of imperial expansion lends another dimension to conquista: it cannot be
imagined *properly* without an interlocutor, also the object of the envisioned expropriation of inhabited, and therefore habitable, territories.[1] Though ceremonies of possession, to borrow Seed's (1995) term,
did take place in the absence of Indigenous interlocutors, the scripts of
these performances were nonsensical without the imagined presence,
at the very least, of the anticipated Indigenous other.[2]

Indigenous inhabitants were not only expected, but indeed desired.
Yet this desire for an Indigenous presence brought to the fore the intended violence of the conquistadors' and missionaries' undertaking.
Staking a claim to land, the literal demarcation of the earth by the
nomos, as discussed at length by Schmitt (2003), is an act of foundational violence underlying the guarantees of the law.[3] For land that is
already inhabited, claims to foundational law must be made otherwise;
the imperial ceremonies of possession must do their best to either erase
the native nomos or incorporate it. At times, paradoxical claims are
made; notably, the Europeans sought unknown lands that were inhabited. The Eurocentrism of the episteme is patent — the only knowledge valued is the European a priori — but the Spanish mode of conquest
also privileges subject and neophyte making in a discourse that swings
between claims of Indigenous lack of knowledge, on the one hand, and
their excessive knowledge, often of undesirable practices, on the other.

The not entirely rational desire of the European venture capitalists
to seek out inhabited terrae incognitae, an obvious contradiction
in terms, set the stage for the overdetermined debates on the nature of
those inhabitants and legalistic inquiries into rights that continue to
receive a priori treatment in any discussion of the "so-called conquests," to recall Las Casas's ironic phrasing: the Papal Donation and

the laws of peoples (jus gentium).[4] It bears repeating that the question of the nature of Indigenous peoples and their rights arose from a specific profit motive from the encomienda exchange: usufruct from the Indigenous in the form of labor and tribute and an increase in the numbers of Christian neophytes. Any discussion of the Papal Donation should begin with this profit motive that depends on contact with (an) other humanity, clearly articulated in Pope Alexander VI's bull *Inter caetera* of 1493:

> Sane accepimus quod vos, qui dudum animo proposueratis aliquas insulas et terras firmas, remotas et incognitas ac per alios hactenus non repertas, querere et invenire, ut illarum incolas et habitatores ad colendum Redemptorem nostrum et fidem Catholicam profitendum reduceretis. (Davenport, 1917, 73)

> [We have indeed learned that you, who for a long time had intended to seek out and discover certain islands and mainlands remote and unknown and not hitherto discovered by others, to the end that you might bring to the worship of our Redeemer and the profession of the Catholic faith their residents and inhabitants.] (76)

Unlike the venture of the bishop of Santiago de Compostela, Diego Gelmírez, discussed in chapter 1, this imagined scenario for material and spiritual profits (i.e., converts to Catholicism) does not equate the losses of the vanquished with the victor's gains. In the bishop's expedition to North Africa, the Muslim infidel remained the enemy from death to enslavement; the infidels' losses were the bishop's gain. In the Alexandrine bull, however, and, it must be added, in other conquistas of Muslim-held territories, the possibility, indeed need, to convert the other complicated the profit motive of absolute hostilities.[5] Instead, it was argued that the inhabitants of the terrae incognitae would have something to gain, that is, Catholicism, despite their losses. In the conversion paradigm, accounting for Indigenous subjects became increasingly complex; by virtue of contact, inhabitants of coveted (yet unknown) lands were both enemies and potential brethren in Christ.

The Eurocentrism of the desire for terrae incognitae raises the obvious question, unknown to whom?, while underscoring a value to habitation that undercuts the self-proclaimed superiority of European knowledge: a Eurocentrism whose axis of being is centered, paradoxically, on habitations external to itself.[6] Clearly, something was to be gained from Indigenous habitation. The potential to profit in spiritual and material terms while making claims that the Indigenous peoples also gained from venture capital's enterprise added a new layer to the discourse of legitimacy, which, as argued in the first chapter, made recourse to two exceptions: the *societas* (moral exception to usury) and free trade and evangelization (geopolitical exception to jus gentium). What made the "win-win" promise of venture capitalism so compelling?[7]

These anticipated interactions in the context of profitable violence redefined tropes of conduct, expected rewards, and unfounded fears. Here we might recall the "writing violence," to borrow Rabasa's (2000) concept, in the *Inter caetera*, which in one breath attested to the existence of peoples who live peacefully in the newly discovered lands (*in quibus quamplurime gentes, pacifice viventes . . . inhabitant*), yet commended the foresight of Columbus and his men in building a well-fortified fort (*unam turrim satis munitam*) whence they could ensure the expansion of the Christian empire and the reach of its enterprise. And yet, as a concept, "writing violence"—the representation of material violence and the symbolic violence capable of reproducing material violence—does not fully encompass the cognitive dissonance created by the discourse of love that permeates the Laws of the Indies and *relaciones*, such as that of Núñez Cabeza de Vaca, even as those laws did little to hide their coercive intent and power. To this end, I would like to repurpose Eugenio Garin's concept of "loving violence" for the Italian humanists' fraught relation with antiquity. According to Garin (1969, 13–14), the humanists could only love antiquity by doing it violence and declaring it dead; only the separation afforded by death allowed for the critical detachment with which humanists used the classics to construct their own identities. For me, loving violence in the Indies serves to conceptualize the intersection of Christian love and profit motive without falling prey to cynicism even as it seeks to analyze the cognitive dissonance; at the same time, it aims to convey the complex transition envisioned between conquistadors and Indigenous subjects as the latter were both enemies and Christian brethren *in potentia*. Similarly, the conversion process, which, in the Augustinian

mold, asks each neophyte to perform some degree of self-annihilation (profession of the former self as dead so that the new Christian may live), demands violence both of the missionary and the convert, even when consent may have been given in the absence of overt physical violence.[8] Finally, loving violence refers to one of the strands of "love" that visualized the wooing of the beloved in warlike conceits, including "conquests," and the waging of war as a violent courtship.

As Greene has argued in *Unrequited Conquests* (1999), the complexity of the desire of European invaders for the Indigenous presence in the Americas may be attributed to a Petrarchan subjectivity expressed as a first-person singular whose unrequited love for a silent beloved (Laura) leads to painful introspection. Yet, unlike Laura, Indigenous subjects of the Americas were anything but unresponsive to the claims and clamor of their invaders, as suggested by the promulgations of Alexander VI and Isabel of Castile. Moreover, when the crown enjoined the conquistadors to treat Indigenous subjects "lovingly" or "sweetly" (*amorosamente, con mucho amor, con dulçura*), this injunction was accompanied by a prerogative to document native beliefs, practices, and social organizations—*to listen*.[9] Paradoxically, these efforts to listen went hand in hand with the efforts to declare dead those parts of Indigenous culture and society that were antithetical to their new life as brethren in Christ.

The *Carta a Luis de Santangel* (1493) inaugurated a "Columbian first-person," the subjectivity of unrequited conquest predicated on desire fueled by asymmetrical relations of power (Greene 1999, 35–76). Beyond unrequited desire, however, I discuss how loving was construed in sixteenth-century Castilian, including sincere yearning for correspondence and consent. By virtue of translation, *eros, storge, agape, philia, cupiditas,* and *caritas* could all be loosely rendered as *amor*.[10] The complexity of "amor" and "amar" as umbrella terms to denote various acts and states of being from passions and emotions to virtues and duties comes to the fore in the entries for *amores* and *amar* by the famous Andalusian lexicographer Sebastián de Covarrubias:

> AMAR, es querer, o apetecer alguna cosa. Amor es el acto de amar, lo primero y principal sea amar a Dios sobre todas las cosas, y al próximo como a ti mesmo. Díxose del verbo Latino *amare*. No tengo que deternerme aquí pues he dado la etimología Latina.

AMORES, de ordinario son los lasciuos, tratar amores, tener amores. Amores, requiebro ordinario. Amoricones, los amores entre villanos. Amorío por amor, término aldeano. Amada la querida. Amigo y amiga, se dice en buena y en mala parte, como amador y amante: amigado el amancebado con la amiga: amigarse, amancebarse. . . . [A]mante el que ama y amantes los que se aman. (2011, 63)

──────

[TO LOVE, is to want, or desire something. Love is the act of loving, which first and foremost is the love for God above all else, and to love one's neighbor as oneself. From the Latin verb, *amare*. I need not explain this further as I have given the Latin etymology.

LOVES, in the lascivious sense, to make love, to have a love. Loves, an ordinary [or vulgar?] compliment. *Amoricones*, the loves among peasants. Love affair (*amorío*) for love, the term used by villagers. Beloved, f., the desired one, f. Friend (m.) and Friend (f.) are used, in a good and a bad sense, as lover and mistress: be-friended [refers to] the man who cohabits with a mistress: to set up house together, to live together. . . . [L]over is the one who loves and lovers are those who love each other.]

Covarrubias opens his entry with an allusion to the appetite for things (cupiditas), which is soon followed by a definition of love that seems to reference Aquinas's understanding of caritas as the *amicitia quae-dam est hominis ad Deum* (friendship of man for God and the love of neighbor), but also as *excellentior . . . omnibus aliis virtutibus* (the most excellent of virtues) (Aquinas 2010, 2-2. q. 23. a. 6).[11] Covarrubias fol-lows the allusion to Aquinas with an elliptical reference to the Latin etymology of *amare* and a more detailed exploration of *amor* in its plu-ral, colloquial forms. Then the *amores* entry explores the ramifications of cohabitation, especially among the lower classes. The *requerimiento de amores*, the forceful courtship of the beloved, belongs as much to lowbrow as to highbrow codes of love in the Romance languages tra-dition, and it is this genre, the declaration of love that seeks, nay, re-quires correspondence, that informs the *Requerimiento* of conquest of 1513, discussed below.[12] The juxtaposition of *cupiditas* and *caritas* in the *amor* entry undergoes a reconciliation via the metalepsis of conquista, one intimately related to translation and the commonplace *translatio imperii, translatio studii.*[13]

As Nebrija (1931, 1) professed, language is the handmaiden of empire. In this case, *amor* as a translation of terms that denote a conflicting affect and ethics—cupiditas and caritas come to mind—was too broad, imprecise, and overarching. It is not that *cupiditas* was never translated into the more precise *codicia*, the desire for things. Moreover, Christians' love for God and for each other (caritas) was the subject of treatises in Castilian.[14] It would also require a stretch of the imagination and an overdetermined faith in the power of language to suggest that all-encompassing *amor* could wreak such havoc in discourses and practices of cupiditas and caritas.

Yet the confusion generated by the metaleptic conflation of love and interest offered a seamless reconciliation, *nominally*, of conflicting passions and virtues. *Amor* was efficient for the purposes of empire, for it reconciled cupiditas and caritas without a change in nomenclature. The synonymous use of amor as caritas and cupiditas offered a seamless transition to an economy of love. However, this metaleptic habitus of conquista, where people and their things and their residences on this earth could become the object of both cupiditas and caritas, was met with resistance by the critics of conquest, for the push to reconcile caritas and cupiditas took aim at Christian values as a whole; the conventional antithesis of *amor* and *interés* was difficult to overcome.[15]

In this sense, Greene's focus on the cupiditas and concupiscence of conquest elaborates one potent aspect of Petrarchan subjectivity that goes to the heart of the confusion of love in the Romance languages in general. What goes missing, however, is the counterpoint of Christian remorse, the literal self-fracturing that characterizes Petrarch's fragmentation of caritas and cupiditas (so memorably rendered in Poem 30 of the *Rime Sparse*, a reworking of the hierarchy of adoration in Psalm 118 in the Vulgate).[16] The subjectivity of venture capital is no less complex than the fragmented self but strives for reconciliation. If Petrarchan subjectivity sees itself reflected on a shattered surface, the expanding subjectivity of venture capitalism (its *scalability*) approaches itself within a labyrinthine world of ramifications and tautologies. While emphasizing the individuality of each partner, the corporate fund believes it has found the shortcut through the eye of the needle.

Its amor for the Indigenous other—that is, caritas and cupiditas, hand in hand—satisfies the body and the soul in their preparations to receive grace and the material comforts of life on earth.

Yet for caritas and cupiditas to gain some kind of rhetorical equivalence, they must have been weighed against each other. The reconciliation was achieved through the logic of market values, weights and measures, in a simulacrum of trade in the public square; in this way, caritas succumbs to cupiditas.[17] It is the only way that Charles V could even postulate the absurd proposition of measuring human life against pearls as a remedy for his confession that the loss of Indigenous life weighed so heavily on his conscience: "Porque estimamos en mucho más, como es rrazón, la conseruaçión de sus vidas, que el ynterese que nos puede venir de las perlas" (Because we value the preservation of their lives much more than the profit that we may gain from the pearls, and rightfully so) (Morales Padrón 2008, 78). But such is the prerogative of the sovereign. Thus Petrarch, who counted himself doubly a sinner for holding Laura higher than treasure and his God's commandments, can't hold a candle to the Holy Roman Emperor's torment in the sixteenth century. In this way, Charles V's conscience is entangled in that most imperial of metaphors: translation.

Contingency Plans as Political Theology

By following the itineraries of conquest designed in the Casa de Contratación in Seville and executed on the islands and Tierra Firme throughout the sixteenth century, let us delve into the contradictions latent in imperial injunctions to love and listen. The capitulaciones signed between the Spanish crown and the commanders of expeditions provided the blueprint for the master fable of conquest.[18] In this master fable, the empire brought civilization and benefits to the lands inhabited by the weaker other: the sovereign showed his or her love for the new subjects by protecting the innocent, freeing peoples from the shackles of bondage, and bringing order to chaos; these others were taught the ways of the good life, including Christianity. At the same time, the empire and its agents received material profits from the lands of others: in kind, in labor, in geopolitical advantage but also in moral value. In the discourse of empire this quid pro quo follows the logic of "love interest," an overdetermined reconciliation of antitheses and unity of purpose among temporal and spiritual sovereigns, subjects, new and old, financiers, and the ever-present enemies and future friends of the state. No longer do cupiditas and caritas struggle for preeminence within the subject as the subject of venture capital and empire has forced a synthesis of the two.

Yet from the juxtaposed items in the contracts, which outline debts and limit obligations, foresee contingencies, and distribute profits, the master fable unravels. The laws seek peace by creating zones of armed conflict; the laws recognize local laws and customs but only to the extent that these do not conflict with the civilizing mission or the profit imperative of the sovereign and his agents. Contracts and laws both underscore and hide anxieties about the moral and material risks at play in each venture, each intended expansion of the empire's reach. Empire accords different values to *love*, *charity*, and *peace* but also labor, commodities, titles, and subjects. Even time—which was considered to belong to no one but God—receives a value.

The Treaty of Tordesillas, signed by John II of Portugal (r. 1481–95) and Isabel of Castile (r. 1474–1504) and Fernando II of Aragon (r. 1497–1516) on June 17, 1494, assigned a value to the spiritual life of sovereigns relative to the church's jurisdiction over spiritual life.[19] In his *Political Theology* (1985, 25), Schmitt's notorious definition of the sovereign as "he who decides on the exception" offers a paradoxical insight into a treaty signed by two temporal sovereigns and adjudicated by the third, spiritual sovereign. How, then, does one enforce rules of an agreement between three "deciders" of the exception? Threats of war and trade sanctions were themselves causes of just war; a treaty made in good faith, to keep the peace, could involve such means of enforcement if they were not themselves infractions of the treaty. In the case of the Treaty of Tordesillas, Pope Alexander VI (r. 1492–1503) made recourse to a spiritual exception in order to force compliance by the two temporal sovereigns of the Iberian Peninsula.

By tying the hands of the church in the realm of the monarchs' consciences, the treaty had a binding effect on the spiritual life of the Iberian sovereigns.

Por mayor seguridad e firmeza de lo susodicho, juraron a Dios e a Santa Maria e a la señal de la Cruz, en que pusieron sus manos derechas, e a las palabras de los Santos Evangelios do quiere que mas largamente son escriptos, en anima de los dichos sus constituyentes, que ellos y cada uno dellos ternan e guardaran e cunpliran todo lo suso dicho, y cada una cosa e parte dello, rrealmente e con efeto, cesante todo fraude, cautela, e engaño, ficçion, e simulaçion, e no lo contradiran en tienpo alguno, ni por alguna manera. So el qual dicho juramento juraron de no pedir absoluçion ni rrelaxaçion del a nuestro muy Santo Padre, ni a otro ningund legado ni prelado que gela pueda dar, e aun que propio motu gela den, no usaran della, antes por esta presente capitulaçion suplican en el dicho nonbre a nuestro muy Santo Padre, que a Su Santidad plega confirmar e aprovar esta dicha capitulaçion, segund en ella se contiene e mandando expedir sobre ello sus bullas a las partes, o a qualquier dellas que las pidieren. (Davenport 1917, 92)

———————

[For greater surety and security in the aforementioned [treaty] they swore to God and Saint Mary and made the sign of the cross and placed their right hands on the Holy Gospels that they would keep and uphold the aforementioned in each and every thing effectively, without fraud, machination, ploys, lies or dissimulations and they will not contradict it at any time or in any manner, and they swore on that oath not to ask for absolution or laxity from our Holy Father or any other legate or prelate who could give it. Even if he were to give it of his own accord, they [John II of Portugal, Isabel of Castile, Fernando II of Aragon] are not to make use of it. By the present treaty they beg of His Holiness that he confirm and approve this treaty and publish his bulls to each party mentioned therein and any other party who may request it.] (99)

By swearing on sacred objects to uphold the terms of the treaty, the three sovereigns formalized the negotiations that had taken place in their absence in the weeks prior to the actual drawing up and signing of the document on June 17, 1494. Yet the possibility that the Catholic monarchs and King John II might foreswear themselves had to be entertained, thus leading to the second oath: should they break the first oath (to uphold the terms of the treaty), the second oath forbade the sovereigns from pursuing absolution from the pope or any other church prelate. Even if the pope or another prelate were to offer absolution of his own accord, the second oath preempted the sinner from receiving it. But what if he or she were to accept absolution? This would have implied a breach of the second oath, and so on. The implied concatenation of oaths, sworn and foresworn, involved a continuous examination of conscience by each sovereign. It also formalized the attempt to create a binding, overarching sovereignty out of this performative, dare I say magical, document.[20] The treaty acted as a barrier to absolution in the event that it was given by any other Catholic prelate; it offered an exception contingent on the other sovereign's exercise of the exception. Yet, like the contradiction inherent in the search for inhabited unknown lands, another paradox emerged from Alexander VI's claim to radical title over the earth.

Paradoxically, the power of the pope to divide the world in two undermined the church's power to intervene in the spiritual terrain at the highest level: the conscience of the sovereign. Or, rather, it attempted to enforce through the threat of passivity, effectively secularizing absolution mechanisms. The church and its prelates intervened in the sovereigns' consciences through nonintervention. If the sovereign indeed resided both within and without the law, such a document, in the area of intersovereign enforcement, entailed an abdication—if only formal and fragmentary—of the church's absolute power to provide succor in the face of spiritual death. The terms of the treaty indicate that for a providential enterprise, ventures in conquista entailed enormous risks, including the increasing fragmentation of corporate institutions in their pursuit of universal sovereignty. The treaty's turn to aporia in the realm of conscience management, however, also calls into question Schmitt's (2003) definition of the sovereign as the *decider* of the exception. In this early modern precursor to the Peace of Westphalia, defined as the absence of war among Christian states (Schmitt 2003, 121), intersovereign sovereignty seems to be performed by the spiritual sovereign's resolve to remain undecided, the pontiff's secular peace.

The pope eliminated absolution as a means for enforcing a treaty between two sovereigns in the matter of exploration and settlement along the dividing line between Portuguese and Spanish domains. This limited appeal to the exception retained absolution only in the matter of adherence to the treaty, nothing more and nothing less. The sharply delimited spiritual no-man's-land served as a guarantee to pursue the spiritual and material profits promised by the three European powers' shared enterprise in the inhabited terrae incognitae of the world. Yet this recourse to the exception, as employed by the highest echelon of the church hierarchy, would be expanded by the confessors in defense of Indigenous subjects less than a decade after the promulgation of the bulls of Alexander VI and Julius I (r. 1503–13) and ratification of the Treaty of Tordesillas.

The conscience of the sovereign was tied to the habitations of the New World. For the critics of the conquista, their refusal to provide absolution to conquistadors became an exception wielded in favor of the oppressed. By decoupling spiritual and material gains, the value of a clean conscience could no longer be written off by the princes of the church. Accounting for sins, for friends and enemies, slaves and enco-mendados, became a habitus codified in law but also an intellectual crutch in the discourse of conquista itself and its unrelenting belief in the transformational power of creative destruction.

This should come as no surprise to those familiar with Derrida's *Politics of Friendship* (2005). As he reminds us in his reading of Montaigne and Aristotle, loving and befriending are actions that create a hierarchy between lover and beloved, friend and befriended, for there is the question of action and passivity or how to make friendship *count*. The politics of friendship is not without its accountants; though "it is possible to love more than one person, Aristotle seems to concede," Derrida *reckons* with the limitations of number in friendship, "to love in number, but not too much so — not too many" (21). Though boundless love, akin to boundless law, seems an impossibility still, Derrida continues,

> it is not the number that is forbidden, nor the more than one, but the numerous, if not the crowd. The measure is given by the act, by the capacity of loving *in act*: for it is not possible to be in act (*energein*), effectively, actively, presently at the heart of this 'numerous' (*pros pollous*), which is more than simple number (*ou gar oión te áma pros pollous energein*). A finite being could not pos-

sibly be present *in act* to too great a number. There is no belonging or friendly community that is *present*, and first *present to itself*, *in act*, without election and without selection. (21)

There is an inherent tension between being and loving in the *present* but accounting for the future. In the conquista, the distribution of love *presently* nonetheless looks to the future *interests* of imperial expansion. Moreover, this love—as we shall see in Charles V's burdens of conscience—trades in death and slavery (i.e., social death) but protests its ability, indeed its duty, to deny the limitations of love and account for new subjects and slaves even if, at *present*, they are enemies to be reckoned with.

The *Requerimiento* (1513)

A short, thousand-word document reviled and rewritten in relaciones, histories, and fiction, the *Requerimiento* was declaimed and performed by armed men and prelates. Its "draconian series of speech acts" re-iterated efforts to "legitimize and perform possession in the terrible union of pen and sword which, nevertheless, veered toward the absurd, even quixotic *avant la lettre*" (Gaylord 1999, 88).[21] The *Requerimiento*'s authors, probably Juan López de Palacios Rubio (1450–1520) and Fernández de Enciso, depict the universal authority of a church whose seat is in Rome *in actu* but, in potentia, could be moved anywhere. As narrated by the *Requerimiento*, the church's universal dominion had allowed its Supreme Pontiff, Alexander VI, to *give* part of the world west of the Cape Verde and Azores Islands to Isabel I of Castile and Fernando II of Aragon and their descendants in perpetuity. The *Requerimiento* was read for the first time in the New World during the expedition of Pedrarias Dávila; in fact, the capitulación stipulated that the new governor take and copy the *Requerimiento* to deposit it with the authorities in Santo Domingo so that Dávila's expedition and others thenceforth could perform the script in all areas to be newly subjugated to the crown.

The origins of the *Requerimiento*'s critique by later commentators can be found within the text itself. In their redaction of this most performative script, the *Requerimiento*'s authors display a self-reflexive understanding of their own limitations, especially in their appreciation for the paradox of the document at hand: it is a "notificaçión y requerimiento que se ha de hazer a los moradores de las yslas e tierra firme del mar Oçeano que *aún no están subjetos* al rey nuestro señor" (notice and requirement that is to be made to the inhabitants of the islands and mainland who *are not yet subject* to our lord the king) (Morales Padrón 2008, 338; emphasis mine). The intended audience is not *yet* subject to the sovereign but will be by the end of the performance; they are but *inhabitants* of islands and a mainland but will become *subjects* by the power of this speech act, this requirement made of them by a representative of the king who is his servant, messenger, and captain. The means for contracting new subjects are imperfect, the text has its messengers acknowledge, but "vos notifico y hago saber como mejor puedo" (I give you notice and will have you know to the best of my ability) (338). Yet this caveat does *not* exempt the *Requerimiento*'s captive audience from the claims, however well founded, of ignorance. What follows is an abbreviated course in Christianity, Western Civilization 101, if you will, that seeks to gloss over the fraught question of (in)vincible ignorance when inviting a new party to undersign a compact.

The *Requerimiento* proceeds with its history of universal descent from "un hombre y una mujer, de quien nosotros y vosotros y todos los hombres del mundo fueron y son descendientes y procreados" (one man and one woman from whom we and you and all the men of the world were and are descendants) (338) and explains how these generations of humanity were dispersed throughout the world, leading to their division into different "kingdoms" and "provinces." It could not be otherwise, "que en una sola no se podrían sostener ni conservar" (for they could not subsist in one area) (338). The original isolation—explained as a necessity for subsistence—will later be expanded by José de Acosta in favor of freedom of movement for trade among nations but in reverse. Writing in the second half of the sixteenth century, Acosta will argue that the dispersion of the human race around the world makes free movement for trade a moral and universal imperative.

Iam vero mercaturae artis hoc propriam est, ut quae apud suos abundant, deferant ad externos, & quibus vicissim illi circunfluunt, reportent suis. Ita enim communis nostri generis autor mortales omnes interse sociandos & quadam communione in officio retinendos existimauit, si sibi essent vicissim opportuni & commodi. (Acosta 1589, 156)

[Furthermore, it is part of the nature of commerce to carry to foreigners what we have much of, and what is superfluous to their needs for them to bring it to us. Thus, it pleased the common Author of mankind for all mortals to associate themselves in this manner and to maintain themselves in unity through mutual communication, so that they might be of mutual help and advantage one to another.] (Acosta 1996, 132)

The free movement for trade doctrine is not further elaborated in the *Requerimiento* but alluded to in the enumeration of benefits promised in return for peaceful subjugation: entry into a neighborhood of new Spanish subjects, like those of other isles (*vecinos de las otras yslas*), who in exchange from the crown have converted and received "privileges," "exemptions," and "gifts" (Morales Padrón 2008, 339). The diversity of habitats and the riches of each land necessitate the free commerce of goods and peoples. At the same time, the common origin of mankind, dispersed, obligates a univocal leadership to Saint Peter, for God "para que de todos los hombres del mundo fuese señor e superior, a quienes todos ovedesçiesen e fuese cabeça de todo el linaje umano donde quier que los hombres viviesen y estuvicsen, y en qualquier lcy, seta o creencia y [dios] diole a todo el mundo por su reyno señorío y juridiçión" (gave him the whole world for his kingdom and jurisdiction so that he would be the overlord of all humankind, to be obeyed by all and to lead the human race wherever humans live and inhabit, and in whichever law, sect, or belief) (338). By this account, local and customary laws (jus gentium) become subject to a universal imperative of greater benefits and progress. Conspicuously absent is Christ himself in this legal formula for universal, Christian empire.

And yet why should Saint Peter's preeminence be so self-evident? Only a few lines earlier the text mentions that humankind could not subsist as one in one place; in this original *historia*, history but also story, the diversity of livelihoods and customs made sense within the divine plan for humankind. That claim to universal dominion rests on an unspoken but necessary analogy to the unifying imperative of free movement for trade. The *Requerimiento* omits Christ from the story, and Christians are just another item in the long list of peoples under the pope's jurisdiction (Moors, Jews, Gentiles, and any other "sect" or "belief"). Moreover, this move toward centralized power in Rome stakes a claim to any place on earth where the pope "pudiese estar y poner su silla" (could reside and place his seat) (Morales Padrón 2008, 339). The list of religions stands in contrast to the list of territories as synecdoche for peoples to which Fernando II of Aragon asserted a claim at the beginning of the document as the tamer of barbarous peoples (*domador de las gentes bárbaras*): Spain, Sicily, Jerusalem, and the islands and mainland of the ocean. The *Requerimiento* does not delve into the fraught tension between spiritual and temporal powers, but the idea of "Christian empire" is conveyed in this contrast between peoples (qua their religions, qua their territories) while promulgating an overarching, universal sovereignty. In this way, the document conveys the universal significance of the Papal Bulls of Donation (*Inter caetera, Eximiae devotionis, Dudum siquidem*) delivered in 1493 by Pope Alexander VI: the papal titles and the monarchy's dominions dovetail in the larger narrative about how the universal jurisdiction and dominion of the church came to be and was being performed through the Spanish crown and its representatives.

At the time of the *Requerimiento*'s composition, the Papal Bulls of Donation were less than two decades old. Though the script for conquest "covers" more than five thousand years, the phrasing of the *Requerimiento* does not acknowledge the relative novelty of the event to the narrative's intended audience (the Indigenous soon-to-be subjects of the Spanish crown). Instead, there is a studied timelessness to it, an effort to avoid pinpointing the year in which papal bulls were promulgated or treaties signed by the two sovereigns on the Iberian Peninsula:

> Uno de los Pontífiçes passados que en lugar deste [San Pedro] suçedió en aquella silla . . . hizo donaçión destas yslas y tierra firme del mar Oçéano a los dichos Rey y Reyna y a sus subçessores en estos reynos . . . segund se contiene en çiertas escripturas que sobre ellos pasaron . . . que podeys ver si quisieredes. (Morales Padrón 2008, 340)

[One of the past pontiffs who took [this one's; i.e., Saint Peter's] place in that seat . . . donated these islands and Mainlands of the Ocean to the King and Queen and their successors in those kingdoms . . . as witnessed by these documents . . . that you may see if you choose.]

The effort to demonstrate the legitimacy of the Papal Donation by showing notarized copies of the documents is worthy of the derision that subsequent commentators have shown it, given the obvious limitations in communication between the crown's representatives and their native interlocutors. What may slip through the cracks of righteous condemnation, however, is the recognition that the *Requerimiento* exemplifies the intent behind Spanish humanists' *poeisis*: the productive use of sacred history, philology, and the written word to compel peoples of radically different traditions and beliefs into consent.

In the *Requerimiento*, one demonstrative pronoun, the difference between "this and that Pontiff" (*este y aquel Pontífice*), separates Alexander VI from Saint Peter, or about fourteen hundred years anno domini. Much is achieved by this abbreviation in time between St. Peter and Alexander VI, almost as if Palacios Rubio and Fernández de Enciso could perform a temporal sleight of hand. Perhaps the authors would have liked the Indigenous interlocutors to believe in its veracity by relegating the Papal Donation to the mythological time of humanity's origins with Adam and Eve or even the pontificate's founding with St. Peter in classical antiquity. It is a rhetorical feat in forced discontinuity, like the narratives of forgeries attributed to Michelangelo and other humanists in the *Lives of Artists* by Giorgio Vasari (1550). Who is that pontiff who made the donation? One of those past pontiffs who succeeded the other one (Peter). When?

Like other Renaissance men eager to pass off their own sculptures as recently excavated treasures to gullible antiquities dealers, the authors readily proffer these documents from some remote, yet so novel, so recent, past and marvel at their own skill, so proximate to the ancient model. The lack of details, the abbreviations, the omissions, the gaps in information—like an arm or a nose broken off a marble figure of a deity no longer venerated or a text truncated midparagraph—only serve to lend the foundational myth and this most recent historical artifact more validity.

Gaining steam from the manufactured antiquities offered, caveat emptor, the reasons for submitting to the aforementioned king and queen are as numerous as all the neighbors who have already submitted to their dominion, following a performance of the *Requerimiento*, just like this one: "con buena voluntad y sin ninguna resistençia" (with good will and without resistance) (Morales Padrón 2008, 338). Yet the rush to sign the deal "without delay" (*sin dilación*) underscores the fragility of the named and unnamed authorities in the proffered narrative. Neighboring inhabitants provide compelling examples of the Christian empire's many benefits. Those who have accepted the missionaries in their midst had been received by their Spanish sovereigns "happily and benignly" (*alegre y benignamente*), so we are told.

The *Requerimiento*'s interlocutors are urged to follow suit, by taking the *necessary, precise,* or *fair amount of time* to understand and deliberate on it (*entenderlo y deliberar sobre ello el tiempo que fuere justo*). The document's insistence on a precise measurement of time to permit meaningful consent incorporates another value—time—into the proposed exchange with their Indigenous interlocutors. The units of *tiempo justo* (fair time) become yet another currency in the moral and material economy of love interest, in the allusions to the dangers of temporal excess. Let us recall that one of the arguments against charging interest (i.e., *lucrum cessans*), analyzed at greater length in chapter 1, was that humanity could not charge for something (time) over which it held no ownership rights. The *Requerimiento* makes no claim to "give" or "bestow" enough time to deliberate the weighty matter of submission to pope and crown. However, as the document alludes to a precise allotment of time for deliberation, just enough to be fair and legal (*justo*), it makes a temporal incursion, fraught with moral peril, into the realm of the *sacer* (sacred). Soon thereafter the *Requerimiento* invokes pursuit of malicious delay (*si en ello dilación maliciosamente pusierdes*) in Indigenous deliberations as a cause for a "just war" (fair, legal, precise). Willing submission ensures incorporation by the sovereigns with "much love and charity" (*todo amor y Caridad*), including the promise of freedom from servitude and recognition of existing property rights as well as privileges, exemptions, and gifts. Conversion is an option, not an obligation, though, as we saw in the earlier clause, the Indigenous subjects are obliged to receive and listen to missionaries. Refusal to submit or malicious delay incurs the full wrath and power at the disposal of the speaker, who threatens to "do to you all the evil and harm that I am capable of" (*vos haré todos los males e daños que pudiere*): all-out war, the full yoke of the church and state, enslavement, and seizure of property (Morales Padrón 2008, 340). Moreover, the fault and the guilt for these malicious actions—defined as such by the speaker, who intends to do the maximum harm and destruction possible—are your own (*sean a vuestra culpa*). In this imagined scene of total and utter devastation, there remains a modicum to be transferred from one party, the Spanish-speaking subject, to his Indigenous interlocutors: guilt and liability.

Decades after his experience in the Darién, Fernández de Oviedo took issue with the legality of the *Requerimiento* in his *Historia general y natural de las Indias,* noting that it presupposed that the Indians had been put in cages, preemptively, for the ideal performance to take place (Fernández de Oviedo 1959, 348). As with most ironic utterances, his statement must be read literally.[22] He also questioned the definition of *tiempo justo*, the legal, fair, necessary amount of time for the Indigenous to make their deliberations following the performance of the *Requerimiento*, noting that the "invitation" to submit to church and crown should be understood in the context of the document's performance, including the motivations of the conquistadors: "para que esté bien con que les convidamos, es menester tiempo porque lo entiendan; y lo que los soldados enseñan es mezclado con propia codicia, y enseñándoles el cuchillo, aunque nuestra voz sea de paz" (for them to take our offer, time is needed for them to understand it; and what the soldiers show and teach is mixed with their own greed, and they show them the knife, though our language is of peace) (Fernández de Oviedo 1959, 348–49). However, Fernández de Oviedo does not go so far as to question the rationale of the (universal) History that served to justify the Papal Donation. Other conquistadors among the Dávila expedition, writing of their experiences in the Darién immediately after, condemned any and all signs of Indigenous resistance.

The sixth man on Guamán Poma's representation of the Spanish Conquest, analyzed in the introduction, is a case in point. Martín Fernández de Enciso, lawyer (*bachiller*), navigator, and author of the *Suma de geografía* (1519), was shocked by the reception that the document, which he had coauthored with López de Palacios Rubios, received in the Darién. Fernández de Enciso (1987, 220–21) reported in the *Suma* that one of the caciques had argued that the pope must have been drunk when he gave away what was not his to give. Though Fernández de Enciso recounts this anecdote to justify the violence used by Dávila and his cohort against the Cenú, the same anecdote would later be quoted by Las Casas to condemn the actions of Dávila and colleagues and underscore the absurdity of the legal script of conquest. The condemnation of the *Requerimiento*, which was ridiculed by its Indigenous interlocutors, relies on a critique of comportment and common sense, not on the basis of legality or philology. The authenticity of the notary's seal, the brand, or the narrative is not in question. It is the belief in the power of one culture's story to subject others to foreign laws and sovereigns that is ridiculed, for only a drunkard would believe such nonsense.

Despite the protestations of the native interlocutors' free will to accept or deny submission to the crown in the *Requerimiento*, with all its attendant consequences, a close reading of the contract signed between Pedrarias Dávila and Fernando II of Aragon reveals the predetermined nature of encomendados and esclavos. As stipulated in the contract for the joint venture, the "choice" the Indigenous had, slavery or encomienda, was predetermined and, thus, a counterfeit of war. Slaves, as Orlando Patterson (1982, 1–16) contends, live the paradox of social death, for that is the life and place of slavery within the social compact. Encomendados, the Indigenous subjects who provided tribute in labor and in goods to the conquistadors in exchange for Christian tutelage, submitted to the slow deaths of their *former* selves in the process of ethno-suicide. The itinerary for war and slave or subject making, as argued below, navigated the "free will" among the residents of neighboring isles with all haste as their fates—slaves or encomendados— were already cast.

THE LAWS OF BURGOS (1512)

The contract signed by Pedrarias Dávila and Fernando II of Aragon "para poblar e paçificar" (to settle and pacify) (Morales Padrón 2008, 89) the Darién offers the mercantile framework within which the *Re-*

querimiento would be performed, and the Leyes de Burgos, passed on December 27, 1512, applied to labor and spiritual exchanges (the encomienda).[23] As alluded to in the earlier section on exceptions to absolution in the Treaty of Tordesillas, critics of the conquista widened the exception to create a blanket denial of absolution as a spiritual bludgeon wielded in favor of the oppressed.

Following the sermon by fray Antonio de Montesinos (ca. 1475–1540), the joint venture in the West Indies between church and state experienced a crisis of legitimacy. The Laws of Burgos were scripted, in part, to respond to this crisis in *fama* (reputation) and in faith. In the sermon, Montesinos condemned the encomienda and the encomenderos to a life without absolution. His refusal, and that of other Dominicans, to give the sacraments of confession and absolution to encomenderos would become the modus operandi for Indigenous advocacy among the religious in the Indies. We do not have the full text of the sermon, only remnants from its gut-wrenching refrain—*soy la voz que clama en el desierto* (I am the voice crying out in the wilderness)—that allowed Montesinos to take on the mantle of John the Baptist. One of the young encomenderos who answered his call to conversion from a life of usury and sin, the future fray Bartolomé de Las Casas, gives a harrowing account of his awakening to the suffering of the Indians in his *Historia de las Indias*. Following Christian precepts on fair exchange values, Montesinos claimed that not only was the integrity of missionary work in the Indies at stake, but also the moral status of the labor and spiritual exchanges (as produced by the encomienda system). Yet, as examined in greater detail in chapter 4, free movement for trade and missionary work was reason enough to justify Spanish ventures in the West Indies. However, since labor, especially manual labor, required remuneration to avoid usury, the crown and the encomenderos had to give something of value in exchange for Indigenous labor. The Laws of Burgos itemize exchange values in greater detail than in any earlier legal code written for the purpose of managing conquista in the Americas.

The Laws of Burgos propose remedies to reform the labor and spiritual exchange in the Americas. As Nemser (2014, 3) has proposed in his reading of the Potosí mines in the Andes and Acosta's *De procuranda*, "The extension of Christian concepts to the colonial space of exception engendered new technologies of pastoral care dedicated to making indigenous bodies live, as well as violent techniques of evange-lization based, as in the silver mines, on letting them die." This colonial space of exception, however, is encoded into law as early as 1512 for the area now known as the Caribbean. The tenth law, for example, urges clergy to provide the sacraments of confession to the Indians and to bury their dead "sin por ello llevar interés alguno" (without profiting from it) (Morales Padrón 2008, 316). The Indians must be taught Catholic doctrine "con mucho amor y dulzura" (with much love and sweetness) (Morales Padrón 2008, 314); encomenderos who failed to uphold their duty to catechize would be fined six *pesos*, or units of gold. These pesos would be distributed equally between the sovereign's trea-sury, the accuser, and the sentencing judge. Other laws regulated times for rest and types of work; for example, pregnant women received a reprieve from manual labor (law 18). Local customs such as the *areitos* (ritual dances) were to be permitted (law 14). Other laws that required the distribution of one hammock per Indian (law 19) and meat on Sundays and other holidays sought to ensure basic living conditions (law 15).[24] However, the crown also required that one-third of all Indi-ans in encomienda serve in the mines, which was, in effect, a death sen-tence (law 25). Moreover, though some instances of jus gentium, such as the areitos, were to be tolerated, others, such as marital and sexual practices and the perceived lack of suitable attire, were not. For ex-ample, Indigenous elites (caciques and their wives) were to dress as be-fitted their station; polygyny and homosexuality were outlawed and punishable by death. That many chose exile in the *monte* (i.e., to live as outlaws) and death over life regulated by the exchanges of the en-comienda system was related in great detail by Las Casas but also al-luded to in Fernando II of Aragon's preamble to the Laws of Burgos. The regent laments the great distances separating the Indians in en-comienda from the encomenderos, a physical distance that needed to be bridged in order to ensure "conversión continua" (continuous and contiguous conversion) (Morales Padrón 2008, 311).

Law 27, however, provides the exception to the labor and spiritual exchange and underscores the tenuous distinction between social life and death in the new colonies. Slaves need not enjoy the benefits of the temporal division between work and rest. The encomenderos had to follow the guidelines for the equitable distribution of time (labor and

rest) and resources (material and spiritual) for their Indians in encomienda *unless* the latter were slaves. This slippage reveals some confusion in the law as to the treatment and identities of Indian slaves and
Indians in encomienda, even if the text of the *Requerimiento*, as analyzed above, would seem to offer a clear option between the two sta-
tuses. Work and rest must be distributed for the indios in encomienda
in keeping with law 27:

> Salvo si los tales indios fueren esclavos, porque a estos tales cada
> uno cuyos fueren los puede tratar como él quisiere, pero manda-
> mos que no sea con aquella riguridad y aspereza que suelen tratar
> a los otros esclavos, sino con mucho amor y blandura para mejor
> inclinarlos en las cosas de nuestra fe. (Morales Padrón 2008, 323)

> [Unless these Indians [brought from other neighboring islands] are
> slaves, because these can be treated as each [encomendero] so
> chooses, but we order that they be treated not with the rigor and
> harshness with which the other slaves are treated, but with much
> love and leniency so as to persuade them in the matters of our
> faith.]

Slavery, or social death, provides the exception to the rules of the encomienda. We can infer that mention of the other slaves refers to the
peoples brought from Africa, and they are the exception to the exception of the encomienda system. Even within the binary of colonial
thought and practice, another Manichaean distinction along the fault
lines of race and geospatial provenance emerges.

The Contract between Pedrarias Dávila and Fernando II of Aragon (1513)

The Laws of Burgos arrived in Santo Domingo with three members of
the Order of St. Jerome in December 1516. They were put into practice,
however, as early as 1513 when Pedrarias Dávila, with fifteen ships and
over two thousand men and women, including Fernández de Oviedo,
set sail for the Darién peninsula in what is now Panama. The contract

between Dávila and Fernando II complicates and in some cases contradicts the performance of choice per the *Requerimiento*'s script.
However, Fernández de Oviedo, who served as undersecretary to Lope
de Conchillos (?–1521) as general notary and slave brander, would
have had to ensure compliance with both the contract and the law. Per
the itemized instructions in the contract, renaming the area Castilla
Aurífera, or Castilla de Oro, from Tierra Firme, itself an erasure of the
native toponymy, is the first order of business (Morales Padrón 2008,
89). What follows is an itinerary that plots how the *Requerimiento* will
be practiced: for enslaving, transporting, and selling slaves; for making
"new subjects" and "pacifying" them; and for settling accounts. A cartography of human habitation receives an itinerary so that the *Requerimiento* may be put into practice. This section is worth quoting at
length as it exemplifies the imbrication of cartography, identity, and
bondage with imperial enterprise, starting with the Spanish sovereign
delineating the stops to be made before reaching the Darién. Fernando II routes Dávila's enterprise accordingly.

> Derrota derecha para la provincia del Darien i sin estorvo ni tar-
> dança del viaje lo pudierdes fazer aveys de tocar en las yslas de los
> Canibales que son isla fuerte Baru San Vernaldo, Santa Crus,
> Gayra, Cartajena, Caramari, Codego que están dados por esclavos
> por razón que comen carne humana y por el mal y dapno que han
> fecho a nuestra gente y por el que fazen a los otros indios de las
> otras yslas y a los otros vasallos y a la gente que destos reynos ave-
> mos enviado a poblar en aquellas partes y por mas justificaçión
> nuestra sy hallardes manera de poderles requerir los requirir que
> vengan a la ovediençia de la iglesia y sean nuestros vasallos y si no
> lo quisieren fazer o no lo pudierdes requirir aveys de tomar todos
> los que pudierdes y enviarlos en vn navio a la ysla Española y allí
> se entreguen a Miguel de Pasamonte nuestro tesorero y a los otros
> nuestros oficiales para que se vendan y el navío que con ellos fuere
> os ha de llevar lo que de la dicha Ysla Española se oviere de llevar
> a la dicha Castilla aurífera y por todas las otras partes que pasardes
> especialmente en qualquier parte que tocardes en la costa de la
> dicha tierra aveys de escusar que en ninguna manera se faga dapño
> a los indios por que no se escandalizen ni alboroten de los christia-

nos antes les hazed muy buena compañía y buen tratamiento porque corra la nueba tierra adentro y con ella vos resçiban y vengan a comunicaros y en conosçimiento de las cosas de la nuestra santa fee católica que es a lo que principalmente os enviamos y deseamos que se açierte. (Morales Padrón 2008, 90)

———————

[Make straight for the Darién without stopping or delaying travel and, if possible, land on the islands of the Cannibals, which are the islands of Baru, San Vernaldo, Santa Crus, Gayra, Cartajena, Caramari, Codego. [The cannibals] are given as slaves because they eat human flesh and for the damages they have done to our people and [the damage] done to the Indians of other islands and the other subjects and people we have sent from our realms[.] If you are able to summon them [*requerirles*] subdue them so that they may obey the Church and be our subjects, but if they do not wish to [obey] or you are unable to subdue them you are to take as many of them as you can and send them in a boat to Hispaniola and hand them over to Miguel de Pasamonte, our treasurer, and to the other officials so that they will be sold[.] Then use that same boat [used for the slave trade of Indigenous people between the Cannibal Islands and Hispaniola] to take whatever [materials, resources] is needed to the aforementioned Golden Castile and in all the other places you pass through. Especially if you are on the coast of that aforementioned land [i.e., Golden Castile] you should take care not to do any damage to the Indians so that they are not shocked and do not riot against the Christians[.] Instead, show companionship and treat them well so that rumors flow inland and with them you shall be [well] received and they will communicate with you and in the knowledge of the matters of our holy Catholic faith which is your primary reason for being sent and we wish you success.]

The instructions given to Dávila seem contradictory and plagued by metalepsis, the confusion of causes for effects and effects for causes. On the one hand, Dávila should take the shortest route possible to the Darién without delay. On the other hand, he must stop at the so-called Cannibal Islands to capture "cannibals," who are supposedly to be taken as slaves for eating human flesh and attacking other Indians and the Spanish king's subjects as well as for loss of life and property. Though the text had already instructed Dávila to enslave the cannibals, it also commanded that the cannibals be given the chance to submit to the crown as part of the performance of the *Requerimento*. These instructions given by the crown prioritize Indigenous consent in order to avoid the propagation of Indigenous insurgencies, understood as responses to the coercive methods prescribed by the law. This emphasis on scandal and irrational uprising speaks to the circular logic of the foundational violence at the heart of this entrepreneurial program, which Las Casas (2005, 112) would later refute in his *Brevísima relación* (1552) with the caustic observation that people raising arms to defend themselves cannot be called rebels if they were never subjects in the first place.

In the preemptive narrative offered by the contract, the slave ship is then to be filled with other cargo and destined for the Darién, which is to be called Golden Castile, and the peoples found there will also be offered the choices of the *Requerimiento* (submission or slavery). There, the Spanish sovereign anticipates their willing submission if his soon-to-be subjects are shown companionship and are well treated by his current subjects, thus making them more amenable to evangelization efforts. At the same time, the sovereign contends that failure to treat these new subjects in a "loving manner" will result in insurgencies and riots (*escándalos o alborotos*). The contract positions the decision-making power of the conquistadors and their Indigenous interlocutors in a sequence of ramifications, functioning like the decision tree in corporate manuals and imperial bureaucracies.[25] However, profit motive is the ultimate horizon for this highly scripted behavior. Finally, the contract foresees the logistics for the transport and sale of the slaves in Hispaniola.

The details of profit distribution among the general and lesser partners of the joint venture immediately follow the itinerary of enslavement. The fourth item specifies that the sovereign should receive the two parts of the booty taken on land and at sea in addition to the regular fifth from the ships that he has outfitted with his capital, that is, "puestos los caxcos" (Morales Padrón 2008, 89). However, in ships outfitted by other investors, the sovereign will receive the ordinary fifth. The distribution of wealth from those ships' agents will follow the customs of booty distribution among the *armada* (land armies) and *marineros* (sailors) (Morales Padrón 2008, 89). The sovereign will provide for the salary of the bishop and clergymen for ten years or once they start tithing the native population, whichever comes first (Morales Padrón 2008, 90–91). The first conquistadors of the Darién (Ojeda, Nicuesa, and Fernández de Enciso) will be made *vecinos* of Castilla de Oro; so, too, would Francisco Pizarro several years later. Then, as another item, among several others, the *Requerimiento* is paraphrased and glossed once more:[26]

> Y en caso que por esta vía no quisieren venir a nuestra ovediençia y se les oviere de fazer guera aveys de mirar que por ninguna cosa se les faga fuera no seyendo ellos los agresores. (Morales Padrón 2008, 92)

> ⸻

> [And if they should not wish to show obedience to us and war must be waged, you must take care that under no circumstances should war be waged unless they are the aggressors.]

Fernando II then paraphrases the *Requerimiento* and broaches the subject of Indigenous ignorance and the stakes involved in their decision making as he accounts for their refusal—hence the logistics of enslavement outlined above—but adds that aggression should be employed *only* as a means of self-defense. He appeals to legal forms, including the new Laws of Burgos that Dávila should make public on his arrival in the Indies.

> Les dareys primero a entender el bien que les verná de ponerse devaxo de nuestra obediencia y mal y dapño y muertes de onbres que les verna de guerra especialmente que los que se tomaren en ella vivos han de ser esclavos y que desto tengan entera noticia y que no pueden pretender ynorançia porque para que lo puedan ser y los christianos los puedan tener sin su sana conçiençia esta todo el fundamento en lo suso dicho. (Morales Padrón 2008, 93)

> ⸻

[First you will have them understand the good that will come to them by submitting to us and the wrongs and damage and deaths of men that will come from war[.] Especially, since those who will be taken alive are destined to be slaves[.] They should be fully informed of this so that they cannot feign ignorance of what they could become[,] and so that Christians may take them with a clear conscience.]

From this passage we can infer that Fernando believes death is preferable to slavery. He also hurries through the fraught doctrine of invincible and vincible ignorances and nescience. Briefly, invincible ignorance in Catholic theology refers to the knowledge that an individual has no way of obtaining and thus provides an exemption from the sin otherwise committed; vincible ignorance refers to a lack of knowledge that any rational person could obtain if they applied themselves to its comprehension. In secular law, the axiom *ignorantia juris non excusat* (ignorance of the law is no excuse) bears some similarity to the doctrine of vincible ignorance. Fernando's haste recalls the value placed on the distinction between tiempo justo and malicious delay in the *Requerimiento*, a distinction that in itself anticipates shared knowledge, customs, and time frames for decision making.

The *Requerimiento*'s summary introduction to Christianity and the Papal Donation would provide fodder for discussions of invincible and vincible ignorances and the relationship between the temporal and spiritual powers of the state and church for years to come. Following the legal formalities, Fernando II of Aragon engages in a candid discussion of the inherent conflict of interest between slave taking and encomienda making, choosing to gloss over the ignorance of his soon-to-be subjects with the more readily vincible ignorance of his actual subjects.

Aveys de estar sobre aviso de una cosa que todos los christianos por que los indios se les encomienden tienen mucha gana que sean de guerra y que no sean de paz y que siempre han de hablar en este propósito y avnque non se pueda escusar de no le platicar con ellos es vien estar avisado desto para el crédito que en ello se les debe dar y pareçe aca que el mas sano pareçer para esto será el del reverendo fray Juan de Quevedo obispo de el Darién y de los clérigos

que están mas sin pasión y con menos esperança de aver dellos yn-
teresse. (Morales Padrón 2008, 93)

———————

[You, [Pedrarias Dávila,] must be forewarned of one thing[:] that
all Christians prefer that the Indians to be given to them in *encomienda* be of the warring and not peaceful [type] and [they] always
speak with this purpose in mind[.] And though not speaking to
them cannot be excused in any way it is good to be forewarned
because we must take their word for it[.] Here it seems that the
most wholesome person for this [judgment] is reverend friar Juan
de Quevedo, bishop of Darién, and other members of the clergy
who are less impassioned and have less hope for the profits to be
gained from this business.]

As in the *Requerimiento*, Fernando II of Aragon makes a sharp distinction between warring and peaceful Indians. However, in this passage of
the contract, he recognizes the interest of the conquistadors in receiving their compensation in the form of "warring" Indians, those destined to become slaves (esclavos), as opposed to "peaceful" Indians,
those destined to become Indians held in encomienda (*indios encomendados*). The encomendados were to submit to the slow deaths of their
former selves in the process of forced labor and the "self-annihilation"
(per O'Gorman) and "ethno-suicide" of confessional evangelization
(per Rabasa); slaves would live the *paradox of social death*, for, as Patterson (1982) contended, such is the life and place of slavery within/
without the social compact, as understood by the slavers. Yet the Spanish sovereign's moment of introspection also reveals a slippage between
encomienda and slavery, for he uses both terms synonymously, a usage
that anticipates, however cynically, modern definitions of war. In the
sovereign's and the entrepreneur's itinerary for war and enslavement or
subject making, they navigated the "free will" among the residents of
neighboring isles with all haste, preemptively, as the others' fates—
slaves or encomendados—were already cast. As noted earlier, Fernández de Oviedo would have been the official responsible for branding
the "warring Indians" once Quevedo had passed his judgment on his
fellow Christians' stories rife with conflicts of interest, stories foretold,
in every sense, in the joint venture between Spanish monarch and
conquistador.

Charles V's Burdens of Conscience and the Contract Signed with Francisco Pizarro (1526)

The preamble to the *Ordenanzas* of 1526 famously refers to Charles V's burdens of conscience (*cargo de conciencia*) that are charged to the unruly greed (*codicia desordenada*) displayed by many of his subjects in the Americas. These phrases are later reiterated in the so-called *Leyes nuevas* (New Laws) of 1542. Though Fernando had not used these catchphrases to refer to the frenzy for gold, pearls, and especially labor in the Americas, he did show concern for conquistadors' interest in making slaves over subjects, hence the recourse to the bishop for his arbitration over the decisions to make war with the consequence for Indigenous enslavement in Golden Castile. In other words, Fernando attempted to organize cupidity or "bridle greed" preemptively: first, by emphasizing the attitudes of love and companionship that Dávila and his cohort were to show to his new subjects in the Darién; and second, by engaging the arbitration of its general partner for each venture — the church — which did not have as much material profit motive in the final status of the Indigenous subjects as it did in their spiritual status as Christians in potentia. In November 1526, Charles V followed his predecessor's example and underscored the role of the church as moral policeman in law and contractual agreements as a counterweight to conquistadors who preferred Indigenous enslavement over holding them in encomienda. As opposed to his grandfather, Charles V placed greater emphasis on Christian stewardship as the mechanism to hold in check the appetites of his unruly subjects and business partners. Yet this greater emphasis on love does not implicate lesser violence, just a more efficient (or *disciplined*, to borrow the terminology of Charles V) use of it.

Remedies are set in place to investigate and punish "culpa de muertes y esclavitud *indebidas*" (guilt for *improper or wrongful* deaths or enslavements) (Morales Padrón 2008, 375), and Charles V turns to the clergy for arbitration on these matters. References to improper or wrongful deaths or slavery would imply that there exist rightful and proper deaths or enslavements, as articulated in the *Requerimiento*. Along the same lines, Charles V's concern for his new Christian subjects (the Indians held in encomienda) is reflected in the injunction prohibiting their return to their old homes "even if they desire it" (*aunque ellos lo quieran*) in order that they be "separated from their vices" (*se aparten de sus vicios*) by living within the encomienda system (Morales Padrón 2008, 376–78). Like Fernando's instructions to Dávila, the sixth and ninth *Ordenanzas* of 1526 paraphrase the *Requerimiento* and its procedure for making new subjects or new slaves. The prohibition against opening new mines similarly allows for an exception, at the discretion of clergy, but stipulates that miners should be treated as "free persons" (laws 9 and 10). However, as made clear in law 11, "new Christians" are subject to the encomienda system and, thus, could be forced to reside near the mines and work on them within the paradigm of encomienda, the labor for catechism exchange.

Charles V's capitulaciones signed with Francisco Pizarro in July 1529 show a concern for proper procedure with regard to creating "disciplined greed." Thus he commends Pizarro, a vecino of Castilla de Oro, and Diego de Almagro, a vecino of Panama, for obtaining the permission of Governor Dávila before leaving for the coast of the Southern Sea in order to conquer, discover, and pacify. Moreover, the costs incurred by Pizarro, Almagro, and their cohort during this first expedition would not be reimbursed by the crown. According to the contract, Pizarro and company spent 30,000 pesos in gold and would continue with the enterprise "con el deseo de nos servir" (with the desire to serve us) (Morales Padrón 2008, 233) but at no expense to the crown. This conquista would be done "a vuestra costa e mission syn que en ningund tiempo seamos obligados a vos pagar ny satisfazer lo que en ellos fizieredes mas de lo que en estas capitulaciones vos fuera otorgado" (at your cost and liability; we are under no obligation to pay or reimburse you for anything more than what is stipulated in this contract) (Morales Padrón 2008, 234). What follows is the famous delimitation of Pizarro's governorship from Tensinpulla to Chincha and items detailing how salaries of officials were to be paid from the distribution of those native lands and labor to be conquered.

Pizarro receives instructions on how to create the positions of mayor, squire, peon, doctor, and pharmacist. Four fortresses are to be constructed for "pacification" purposes, at the conquistadors' expense. In fact, the document goes to great lengths to specify that neither

Charles V nor his heirs are obligated to pay for construction or upkeep of these fortresses. However, the crown does provide funds for artillery and ammunition, which would be disbursed at the Casa de Contratación in Seville. Charles V includes exemption from some tariffs (*alcabalas*) on imports and exports for a ten-year period as an incentive to the conquistadors. Moreover, the ordinary fifth will apply to all wealth gained from mines, trade, and mounted raids (*minas, rescates y cabalgadas*) (Morale Padrón 2008, 235). Some gifts (*mercedes*) bestowed on the conquering party include titles of lesser nobility (*hidalgos de solar conocido*) and keeping their rights to land and labor in Castilla de Oro (or Castilla Aurífera) or to sell them if they so choose (Morales Padrón 2008, 236–37).

Yet there are some items that concern my larger discussion of enslavement and encomienda. Item 19 refers to the crown's gift of African slaves who are "free of all rights" to be traded in the Caribbean en route to Peru. The crown will deduct their worth from its own treasury.

Otrosy vos daremos licencia como por la presente vos la damos para que destos nuestros Reynos o del Reynos de Portugal e yslas de cabo verde o de donde vos o quien vuestro poder oviere quisieredes e por bien tovieredes podays pasar e paseys a la dicha tierra de vuestra governación cinquenta esclavos negros en que aya a lo menos el tercio hembras libres de todos derechos A nos pertenescientes con tanto que si los dexarades todos o partes dellos en las yslas española san Juan y cuba e Santiago o en castilla del oro o en otra parte alguna los que dellos ansy dexaredes sean perdidos e aplicados e por la presente los aplicamos a nuestra camara e fisco. (Morales Padrón 2008, 238)

[By this document we give you license so that you, or whoever has your power of attorney, may take fifty slaves free of all rights (of which at least one-third will be women) to the lands of your governorship [Peru.] These [slaves] belong to us and may come from our realms or the realms of Portugal and the Cape Verde Islands or wherever you wish. If you leave them all or some on the islands of Hispaniola, San Juan, Cuba and Santiago or in Castilla del Oro or somewhere else they will be debited from our own account and treasury.]

Charles V treats the slavery of Africans as a gift that can be written off with the precision of an accountant ensuring that his books are in order. Item 25 then makes references to the 1526 *Ordenanzas* for procedures regarding the Indians and the encomiendas.

Despite the concerns for the sovereign's conscience elaborated in these *Ordenanzas*, the contract between Charles V and Pizarro and company, which was written within their legal framework, hardly makes any reference to the crown's new subjects. This may be due to the sovereign's explicit approval of Pizarro's modus operandi in the initial venture and his experience on the Isla del Gallo. Unlike Cortés, who had no initial contract with Fernando II of Aragon or Charles V, neither Pizarro nor his brothers nor Almagro had shown any signs of insubordination or intent to commit regicide. As vecinos of good standing in Castilla de Oro, they all had experience with the *Requerimiento*. The contract's main concern was to organize expectations, profit margins, and motives in greater detail and refer back to Indigenous labor regulation as a framework with which all parties seemed to be largely familiar. Unlike Fernando II of Aragon's contract with Dávila, which had to include mechanisms for introducing new law into the colony as part of the venture agreement, this contract largely took most items related to Indigenous peoples, especially the script of the *Requerimiento*, for granted.

THE CONSCIENCE OF THE SOVEREIGN

Charles V's preamble to the New Laws, which were promulgated on November 20, 1542, condemns unruly greed (*codicia desordenada*) as the root of the violent excesses employed by the king's subjects against the Indians. Lamenting, as he had earlier, in the *Ordenanzas* of 1526, that the violence in the Indies weighed heavily on his conscience, the sovereign proposed what was to his mind a measured response to the pall of vice that had shrouded the Spanish Empire. Weighing in on his conflicting motivations for incursions into the Indies, Charles V displayed his reasoning for the new prohibitions on certain economic activities. However, by the 1540s, the conscience of the sovereign had fully assimilated the cost-benefit analysis of a moral economy where love and interest dovetailed.

Law 24 is a case in point for its emphasis on moral efficiency. This law requires the immediate cessation of pearl fishing *"si* les paresçiere que no se puede escussar a los dichos yndios y negros el peligro de la muerte" (*if* it is believed that the aforementioned Indians and Blacks cannot be excused from the risk of death) (Morales Padrón 2008, 435; emphasis mine). Charles V evaluates the almost certain death of subjects, on the one hand, and the loss of profit, on the other: "porque *estimamos en mucho más,* como es rrazón, la conseruaçión de sus vidas, que el ynterese que nos puede venir de las perlas" (because *we place greater value*, as is reasonable, on the conservation of their lives than on the interest that we could gain from the pearls) (435; emphasis mine). In addition to the cessation of pearling operations, the reforms attempted to ameliorate living conditions for the sovereign's Indigenous subjects, eliminate corruption in the governing bodies of the Indies overseas (Audiencias), and streamline judicial review of criminal and civil cases by the Council of the Indies in Seville, the highest governing body over the Indies with executive, legislative, and judicial powers. However, the New Laws are most famous for their reorganization of labor-catechism exchanges, or encomienda, and slavery in the Indies.

The encomienda, whereby the Indigenous gave labor in exchange for religious stewardship, had been condemned for its abuses ever since fray Antonio de Montesinos had inveighed against the institution in his sermon on the fourth Sunday of Advent in 1511. Yet the New Laws did not abolish slavery; rather they issued guidelines for remedying *illegitimate* cases of past, Indigenous enslavement (laws 20 and 22). Similarly, the New Laws' reform of the encomienda system tried to reduce abuses against the Indigenous *and* redress inequities in remuneration among the earliest conquistadors. Thus, law 17 redistributed Indians from those encomiendas that had an excess number of them to the first conquistadors (*primeros conquistadores*) who had none, colonists who were married, and, ultimately, the crown. This new law directly contradicted the capitulaciones between the crown and conquistadors, which had authorized the labor for religious stewardship exchange to the conquistadors and their descendants, the so-called encomienda in perpetuity. Less than one year later, following the outbreak of rebellion by the encomenderos in the Viceroyalty of Peru, the

prince (and future monarch Philip II) would overturn the newest reform of the encomiendas, for which Father Las Casas would famously
take him to task in the *Brevísima relación*. The rebellions of the encomenderos also highlighted the tensions that were involved in making
the full transition from a series of joint ventures among private and
state actors to the sovereign's direct stewardship of the religious and
economic enterprise of empire.

As explored above, the examples from the New Laws bring to the
fore the contradictions of this moral economy in which labor and conscience had exchange values. The union of the two antithetical poles of
amor — caritas and cupiditas — was an unhappy one. Yet by the end of
the sixteenth century, they coexisted in a double-entry bookkeeping
system, seemingly without contradiction, in the *Ordenanzas* of 1573.
The question remains, how to reckon with the conquest when souls
and bodies have exchange values with pearls and other tangible, itemized commodities? By these *Ordenanzas*, accounting for sins and
souls, gold and precious gems, and slaves and Indians held in encomienda was intrinsic to the habitus of conquista and its metalepsis.

ORDENANZAS OF 1573

As Todorov (1984) observed, the 1573 *Ordenanzas* prescribe dissimulation for initial encounters between the conquistadors and the inhabitants of terrae incognitae. What remains to be seen, however, is whether
this order to dissemble also means that the crown's pursuit of "love,"
"charity," and "peace" toward its American subjects or, to be more precise, American subjects in the process of becoming was any less sincere.
The crown's declarations of love, peace, and charity should be taken at
face value; that is, these affects existed in relation to other commodities
within the Christian and imperial system of exchange values. In this
sense, the declarations of love were no more a semblance than the Fuggers' request for a papal pronouncement on the moral validity of the
triple contract. By the time the 1573 *Ordenanzas* were promulgated,
reckoning with sin and material wealth was a balancing act performed
on an imperial scale with utmost sincerity.

This does not mean that the moral economy of empire was not met with resistance. As Negri (2008) has argued, Macchiavelli prescribed dissimulation to the embattled, self-made leader in response to the power of the multitude. It is but one of the options because the "multitude's unity of action is the multiplicity of actions it is capable of" (67). In response to the increasing infamy surrounding the so-called conquests in the mid-sixteenth century, the grammar of imperial law prescribes a series of discursive moves made in earnest, including dissembling, not only to address insurgency but also to paper over the widening fissures between interest and love created by their systematic use as synonyms one generation after the 1542 New Laws. The union of cupiditas and caritas in loving empire spawned circumlocutions in legal discourse to widen the boundaries of the nomos, to make it all-encompassing even as it sought to mask its agents' use of force.

There are hundreds of ordinances, often couched in the conditional, attempting to anticipate, overcome, and erase insurgency. The decision tree aspect of contracts between sovereign and conquistador, as we have seen in the Dávila contract, is translated back into law, branching and multiplying in ramifications. Why question the validity of the crown's "burdens of conscience"? Merely indicating that these burdens did, indeed, have a *value*, which generated new burdens, reveals a simulacrum of earlier injunctions against usury: the unnatural, self-replicating specie of actions and their consequences that bore little resemblance to the proverbial fruit-bearing tree beyond its uncanny ability to self-replicate and reproduce. Let us continue to take the empire at its word, then, and examine how moral risks could be mitigated by a change in name and a renewal, however fraught, of loving discourse.

The 1573 *Ordenanzas* show little appetite for risk, in moral or material terms. For the crown, the pecuniary stakes in play changed with said *Ordenanzas*, although the moral and material stakes could not have been higher. Although the crown's contributions in capital to the enterprises in the Americas had never been overwhelming, law 25 declared that the crown would no longer participate with any capital investments whatsoever. It also made the distribution of wealth—in lands, grazing rights, and Indigenous labor—in new settlements proportional to the amount of the original investment (law 47). Similarly, the crown renewed its provisions for receiving "carried interest" (20 percent, or one-fifth) from all mining and pearling operations (law 50).

These laws reaffirmed the practice of profit distribution seen in the contracts drawn up by the Spanish crown with Dávila and, later, Pizarro: namely, land, labor, and nobility titles would be commensurate to the individual's original capital investment.

Moreover, the *Requerimiento*, or something similar, continued to be performed, as law 13 stipulates:

> las personas que fueren a descubrimientos por mar o por tierra tomen posesión en nuestro nombre de todas las tierras de las prouincias y [Tachado: tierras que descubrieren] partes adonde llegaren y saltaren en tierra aziendo la solenidad y autos necesarios de los quales trayan fee y testimonio en pública forma en manera que haga fee. (Morales Padrón 2008, 490)

> _____

> [The persons who are to make discoveries by land or sea should take possession, in our name, of the lands and provinces and [struck through: lands to be discovered] areas where they may arrive and land. They shall make the solemn and necessary acts to which they will give faith and testimony in a public manner.]

"Lands to be discovered" (*tierras que descubrieren*) was the phrase that reiterated and highlighted the contingency of "the discoveries" on the "solemn acts" and, unsurprisingly, became the target of attempted erasure. Perhaps this was due to the embarrassment at the self-fulfilling, performative acts of possession; yet the circumlocution, "areas where they arrive and land," nonetheless describes the tautology of the acts and brings greater attention to the details of performances that were already learned, an ingrained habitus of conquest. Similarly, the encomienda continues to function and forced resettlements are reaffirmed as the norm (laws 50 and 58). Though the laws repeat concern for Indigenous subjects within the framework of divine providence, prescribed reactions are contingent on resistance and insurgency.

The Indigenous are approached as "friends" but treated as enemies; the public square is to be cordoned off in order to wage preemptive attacks against the Indigenous (law 113).[27] Law 136 revisits the modality of the *Requerimiento* and its enumeration of benefits as part of the peaceful submission package. The law is expressed in the conditional construction.

Si los naturales se quisieren poner en defender la poblaçión se les dé a entender como se quiere poblar allí no para hazerles algún mal ni tomarles sus haziendas sino por tomar amystad con ellos y enseñarlos a bivir políticamente y mostrarles a conocer a dios y enseñarles su ley por la qual se salbarán dándoseles a entender por medio de los religiossos y clérigos y personas que para ellos diputare el gouernador y por buenas lenguas y procurando por todos los buenos medios posibles que la poblaçión se haga con su paz y consentimiento y *si* todavía no lo consintieren hauiéndoles requerido por los dichos medios diuersas vezes los pobladores hagan su poblaçión sin tomar de lo que fuere particular de los indios y *sin hazerles mas daño del que fuere menester* para defensa de los pobladores y para que la poblaçión no se estorue. (Morales Padrón 2008, 515; emphasis mine)

[*If* the natives were to mount a defense against the [new] settlement make them understand that we wish to settle there. [We do not intend] to harm them or take their property but to befriend them and teach them how to live the good life and show them to know god and his law through which they will be saved by the religious and clergy and persons and good interpreters to whom the governor has delegated this mission. By all good means possible [procure] that the settlement be made with their peaceful consent. And *if* they still do not consent, even if they still have not consented by different means on various occasions, the settlers should settle without taking what belongs to the Indians and *without doing more harm that what is necessary* for the defense of the settlers, so that the settlement is not impeded.]

Unlike the earliest version of the *Requerimiento*, this legal document does not detail the painful consequences entailed in any refusal to submit; rather, it emphasizes what the natives stand to gain by consent. The litany of the empire's benefits expands the "love interest" of conquista via an economy of scale, in the metalepsis of contiguity and similitude. By virtue of rhetorical largesse, it seeks to erase the violence of conquista and its ignominy but to retain the habitus of love interest.

The circumlocution of violence, and its attempted erasure, only serves to delineate it, much like a palisade constructed around a public square. No longer can the conquistadors take the property of the Indigenous inhabitants, according to the letter of the law. How can they stand to make a profit? Necessity and defense mitigate against the main benefits of the societas and its practitioners: shared risk and ownership. As proposed in the first chapter, the great promise of venture capital is not collateral offered as security or the interest garnered on repayment of a loan. The great profitability of carried interest resides in its continued appropriation for posterity. So the law orders a paradox: appropriation in a property vacuum but appropriation nonetheless.

The law thus depends on two fictions: the settlement of inhabited lands without misappropriation and the absence of aggression. In this imagined no-man's-land, there are no aggressors; both the Indians and the settlers are "defenders." Even so, measures of harm may be doled out to would-be friends. Once the land is pacified (*estando la tierra pacífica*), the Indians are to be distributed in encomiendas or repartimientos (redistributions, another redundancy) and are obligated to pay "moderate tribute" in kind (usufruct, i.e., fruits of the earth). Peaceful evangelization and pacification include taking hostages under the premise of offering an education in "proper attire" (law 142). Yet the law concedes its ignorance on all the manners to proceed "conveniently" and leaves other means necessary (*otros medios que paresçieren conuinientes*) to the discretion of the "pacifiers," "explorers," and "discoverers" (conquistadors in everything but name).

> *Si* para que mejor se paçifiquen los naturales fueren [*sic*] menester conçederles ynmunidad de que no paguen tributos por algún tiempo se les conçeda y otros preuillegios y exençiones y lo que se les prometiere se les cumpla. (Morales Padrón 2008, 518; emphasis mine)

> [*If* it is better to concede a temporal immunity from paying tribute in order to procure the pacification of the natives, do this and grant other privileges and exemptions and ensure that everything promised is fulfilled.]

Contingency plans in the law thus prescribe exemptions to the law and include intelligence-gathering measures that will lead to further ramifications in the decision tree. The circumlocutions of the law must concede that it has boundaries. Paradoxically, the excess in codifying the love interests of empire only served to highlight the law's limitations. "For to love friendship," as Derrida (2005, 29) contends in addressing Nietzsche, "it is not enough to know how to bear the other in mourning; one must love the future. And there is no more just category for the future than that of the 'perhaps.'" Once the contracts underwriting the enterprises and the laws regulating these enterprises are juxtaposed, a series of paradoxes emerge: "loving empire" make claims to providential discovery while engaged in a never-ending pursuit of contingencies; the apologists for Spain's empire insist on providence as the telos in discovery, conquest, exploration, yet the imperial apparatus codified its activities in the Indies in response to the happenstance of friendship and enmity in all earnestness.

CHAPTER THREE

Telling Islands in the *Claribalte* and the *Historia de las Indias*

The word *order* recurs obsessively in the instructions imparted in 1513 by the King, or rather by his council of advisers, to Pedrarias Dávila, leader of a Spanish expedition that pushed beyond the conquistadors' original foothold in the Caribbean once accommodation to the New World environment had readied them for further violent expansion and colonization.

—Angel Rama, *The Lettered City*

I do not doubt that if the Admiral had believed that such a pernicious event would occur as it did, and had he known the first and second conclusions of natural and divine law, as he knew of cosmography and other human knowledge, then he would have never dared to introduce or initiate anything that would wreak such calamitous damages, because nobody can deny that he was a good and Christian man.

—Bartolomé de Las Casas, *Historia de las Indias*

In and around 1515, two letrados faced turning points in their New World careers on the island of Hispaniola: Gonzalo Fernández de Oviedo and Bartolomé de Las Casas. The former arrived in Hispaniola already a bureaucrat in the service of the Dávila expedition as notary, comptroller of the gold foundry, and enslaver of Indians in the Darién; the latter was about to begin his career as a writer and advocate in the service of the Dominican order and the crown, on behalf of the

Indigenous peoples of America, after living on Hispaniola and Cuba for more than a decade. After serving as a chaplain to conquests on the island of Cuba and suffering a crisis of conscience, Las Casas abandoned his encomienda and enslaved Indians in Cibao, Hispaniola, in 1515, in order to advocate on Indians' behalf before Fernando II of Aragon in Spain. Las Casas had arrived in Hispaniola in 1502 and had participated in a number of expeditions, including one spearheaded by Nicolás de Ovando and another by Pánfilo de Narvaez in Cuba (Wagner and Parish 1967, 13–20). His conversion to advocate for the Indians is marked in the written record by his penning in 1516 of the *Memorial de remedios para las Indias* (Las Casas 1995, 23–48); in this first *Memorial*, one of the recommendations he makes to the crown in order to ameliorate the working and living conditions in Indigenous communities is to grant them licenses to have white and black slaves brought from Castile ("dándoles Su Alteza licençia para ellos y haciéndoles merçed de que puedan tener esclavos negros y blancos, que los pueden llevar de Castilla") (Las Casas 1995, 36). Another *Memorial de remedios para las Indias* (1518), which is notorious for promoting the replacement of Indigenous labor with that of enslaved Africans in the Indies, was his second written foray into imperial policy making and criticism (Las Casas 1995, 49–53); this first implicitly held position on the Portuguese conquests in Africa he later retracted and lamented in his *Historia de las Indias* (1527–61). If Las Casas spent the rest of his life revisiting his deeds and writings from his earliest experiences in the Caribbean, his similarly prolific counterpart, Fernández de Oviedo, acted conversely, attempting to erase his experience as slave brander in the Darién, even as he leaned on his other official duties as comptroller and notary in the same expedition to give an account of it with the authority of eyewitness testimony in his *Sumario* (1526) and monumental *Historia general y natural de las Indias* (1535).

In this chapter, the juxtaposition of two "minor" works—or a minor episode within a larger work, as in the case of Las Casas—by two writers who were foundational in the field of Latin American colonial letters allows us to visualize the differences between writing imperial enterprise in a metaphoric versus a metonymic vein. I contend that islands, and the narrative archipelagoes constructed by these authors, serve both letrados to construe their place as moral actors within the larger arc of imperial history. In both narratives, chivalric fiction and history, the vulnerability of characters on islands—especially those waylaid by shipwreck—places contingency for individual persons in sharp relief while calling into question the mechanisms for action implemented by imperial, corporate personhood within the metaleptic habitus of providential thinking that seeks to isolate disaster as happenstance.

THE LETRADOS' TELLING ISLANDS

As he waited in Santo Domingo for a berth on a ship for his return trip to his native Spain, Gonzalo Fernández de Oviedo wrote the first work of fiction in the New World. Prior to this brief respite on the island of Hispaniola, Fernández de Oviedo had been employed as a letrado, an official with writing duties, in the expedition of Pedrarias Dávila in the Darién, whose governing documents—the contract and the *Requerimiento*—were analyzed at length in the previous chapter. In the prologue to this chivalric novel known as the *Claribalte* (1519), Fernández de Oviedo made much of one of his duties as a letrado in this expedition "estando yo en la India y postrera parte occidental que al presente se sabe, donde fui por veedor de las fundiciones del oro por mandado e official del catholico rey don Fernando" (when I was in the Indies, at its westernmost point explored to the present, where I was the comptroller of the gold foundry by order of King Fernando) (2002, 53). In the same prologue, he failed to mention the other duties he had performed as undersecretary to Lope de Conchillos during the Dávila expedition.

A copy of the contract Fernández de Oviedo had signed with Conchillos before their departure from Spain lists the other duties he performed on Tierra Firme. These included overseeing and certifying the "fundición e mareación, e escribanía de minas, e del crimen e juzgado, e el oficio del hierro de los esclavos" (foundry and navigation, the notarization of mines, and the office of slave brander) (quoted in Pérez de Tudela Bueso 1957, 418). As discussed in chapter 2, the enslavement of Indigenous persons was anticipated in both the contract signed by Dávila and Fernando II of Aragon and the *Requerimiento*, first brought to the West Indies and Tierra Firme with the Dávila expedition.

As the slave brander and scribe on the expedition, Fernández de Oviedo would have had to certify compliance with both governing documents, even if the differences between a contract between specific persons for the formation of a corporate personhood, on the one hand, and a generic script, on the other, could generate conflicting expectations for behavior, both by the Indigenous and by the limited and general partners involved in the Dávila expedition. Considering all the various hats Fernández de Oviedo wore on this expedition, he personifies the dictum Rama (1996, 22) famously employed to describe the social relations of the letrados in the New World: "servants of power in one sense, the *letrados* became masters of power, in another." It was his official capacity as slave brander, as *master* over the lives of others, however, that Fernández de Oviedo attempted to erase from his first work as a published author, when he penned his prologue to the *Claribalte*, and later from the historical record as a whole, when he wrote his *Historia* and the *Batallas y quincagenas*, now as a writer of greater renown in the service of the crown.

Many years later, when he wrote his *Historia*, for which he is better known, Fernández de Oviedo (1959, 67) continued to gloss over his duties as a slave brander in the Darién, underscoring instead that he "llev[ó] a cargo la escribanía general del secretario Lope Conchillos y el oficio de la fundición, allende del que yo me tenia de veedor" (performed the duties of general notary for the secretary Lope Conchillos, and officer of the foundry, of which I was the comptroller). As both servant and master of power, the self-appointed historian to the crown continued to erase his mastery over the Indigenous bodies whom he had branded as slaves decades earlier. If, as seen in chapter 1, the ideal merchant could reconcile the accounts written under various temporal horizons (daily, quarterly, and yearly) in order to ensure *credit*—credibility and, thus, creditworthiness—the imperial letrado was similarly tasked, though with the added complication of reconciling conflicting instructions and interests, including his own, as was the case in the treatment of Indigenous persons, outlined both in the expedition's contract and in the *Requerimiento*. Rather than provide an account for his activities as slave brander, Fernández de Oviedo preferred to erase that performance from his record.

On his earliest expedition, as slave brander, scribe, and comptroller of the gold foundry, Fernández de Oviedo performed the duties of the lettered bureaucrat that were instrumental for what Saldaña (2016, 23–24) has called the "temporal, legal, and psychic mappings" that were foundational in the "spatial production of the racial geographies" of the Americas. This included certifying the reading of the *Requerimiento*, whose reiterated performance in the West Indies exemplifies what Butler (1990, 191) would consider the performativity of gender and identity through the exterior "stylized repetition of acts." Through the reiteration of the binary choice imposed on Indigenous interlocutors—envisioned as male father figures—the opposition between "good" Indians (willing subjects of the crown, Christians in potentia) and "bad" Indians (insurgents, enslaved, or the dead in potentia) was institutionalized. Yet the institutions performed by conquering bureaucrats were not without their contradictions.

Whereas the contract Fernández de Oviedo signed with Fernando II of Aragon preemptively mapped the future of the enslaved or neophyte Indians, the *Requerimiento* ostensibly offered choices for Indigenous interlocutors to articulate their will (though well within the binary option of subject or slave). As the official responsible for branding Indigenous slaves and certifying compliance with the *Requerimiento*, Fernández de Oviedo would have had to settle the contradictory claims and requirements of two competing narratives: on the one hand, accounting for Indigenous subjecthood or enslavement, as prescribed in the contract; on the other, certifying the reiterated performance of choice, as laid out in the *Requerimiento*. In this way, though the instructions given to Dávila and his men betray a compulsion for order, as famously observed by Rama (1996), the lettered officer of the crown would have been responsible for validating and reconciling—or at the very least glossing over—these contradictory instructions in subsequent accounts made in an official capacity. In the face of such a daunting task, Fernández de Oviedo favored erasure over narrative consistency and chivalric fiction over the historical and ethnographic writing, preferences that would characterize not only his own production but also the field of colonial Latin American letters as a whole.[1]

The *Claribalte*, which adheres to the conventions of the chivalric novel, seems hardly worthy of note beyond its entirely circumstantial status as the first work of fiction in the field of colonial Latin American letters, which has been defined, notoriously, by the absence of fictional writing in the sense of imaginative, fantasy writing.[2] And yet, though I wholly subscribe to evaluations of the novel's mediocrity, as elaborated by many scholars, including Gerbi (1949), it is precisely its mediocrity, in the etymological sense, within the conventions of Iberian chivalric fiction that strikes me as extraordinary, once we bring the context of its composition into consideration: it was the first work written by a letrado following his daily experience with human life and death as produced through the legal fictions of the conquest, of which he was an officiant.[3] Written by an author who earned his living as a letrado, the novel seems to turn its back on the vantage point of hindsight as a rest area for Fernández de Oviedo to reflect on his recent experience in the Darién. Instead, the first fiction of the New World, penned during Fernández de Oviedo's intermezzo on an island, betrays an author wholly intent not only on erasing his immediate experience in the New World, but on glossing over the New World itself, especially its peoples, from the written record.

The second section of this chapter is devoted to another narrative arc, produced in this case by Bartolomé de Las Casas, this one stringing together the conquests of different islands, from the Caribbean to the Atlantic islands to Hispaniola again. I analyze the anecdote about a plague of rabbits on the island of Madeira, as told by Las Casas, within the larger narrative he tells about the conquest of America but also of Africa and Goa and, of course, Hispaniola, and the moral trajectory he constructs for Christopher Columbus both in the *Historia de las Indias* and as his editor of the *Diario a bordo* (ca. 1530). Indeed, the marginal notes by Las Casas in his copy of the *Diario a bordo* — itself a summary and a copy of the original Columbus gave to Isabel of Castile after his first voyage — serve not only to correct or comment on the logbook but also, as Ruhstaller (1992) has argued, as notes for his redaction and narration of the sections on Columbus and his voyages and what they signified for the Spanish Empire as a whole in the *Historia de las Indias*.

In his representation of Columbus and his role in bringing Christianity to the West Indies, Las Casas struggles with contingency and providence, as if navigating between Scylla and Charybdis. Written over thirty years, the *Historia de las Indias* was not published in his lifetime and suffered many transformations and editorial changes as Las Casas struggled with telling the story of the conquest right, from the *beginning*, though that task became increasingly difficult. Where and how to begin? The similarly monumental *Apologética historia sumaria* (1536), for example, was an offshoot of the original *Historia de las Indias* project. In the *Apologética*, Las Casas compared the cultures and civilizations of pre-1492 America with the pagan Greeks and Romans in order to complement the theological arguments he had made in *De unico vocationis modo* (1537). In this theological treatise, which he attempted to put into practice in Vera Paz, Las Casas made the argument that the only valid model for missionary work was the performance of Christ's apostles, that is, peaceful and at the mercy of Indigenous peoples. Las Casas, thus, served as a letrado while increasingly questioning the bases of the Spanish Empire's legitimacy.

For Las Casas, who in addition to writing histories and theological treatises also served as the editor of Columbus's ship log of his first voyage (1492–93), the difficulty of reconciling the episodic nature of the diary's daily entries, and the individual choices documented therein, with Columbus's figure in "universal history" would lead him to write history in the subjunctive mode. The counterfactual overrides the factual in the *Historia de las Indias*, leading him to propose in this account of the encounters on or shortly after October 12, 1492:

> Yo no dudo que si el Almirante creyera que había de suceder tan perniciosa jactura como sucedió, y supiera tanto de las conclusiones primeras y segundas del derecho natural y divino, como supo de cosmografía y de otras doctrinas humanas, que él nunca osara introducir ni principiar cosa que había de acarrear tan calamitosos daños, porque nadie podia negar él ser hombre bueno y Cristiano. (Las Casas 1994, 560)

> [I do not doubt that if the admiral had believed that such a pernicious event would occur as it did occur, and had he known the first and second conclusions of natural and divine law, as he knew of cosmography and other human knowledge, then he would have never dared to introduce or initiate anything that would wreak such calamitous damages, because nobody can deny that he was a good and Christian man.]

The burden of the first clause becomes increasingly salient to Las Casas following the publication of the *Brevísima relación* in 1552. Unlike Columbus, who could ostensibly claim ignorance of the consequences of his action before setting sail from Palos, the advocacy and publications of Las Casas and his followers in subsequent decades had made it increasingly difficult to claim ignorance, at least not of the "invincible" kind. This realization led Las Casas to conclude, shortly before his death, that Columbus's burdens of conscience had been carried by Spain as a whole, obligating the Spanish monarch and his subjects to abandon the Indies.

The Knight of Good Fortune and the Slave Brander

The *Claribalte*'s obscurity is only overshadowed by the relative disregard shown of Indigenous enslavement in the Spanish American colonies until recently. However, in their recently published monographs, Andrés Reséndez (2016), Nancy Van Deusen (2015), and Dan Nemser (2014) have shed light on the widespread practice of Indigenous enslavement and the acts of Indigenous resistance to their bondage, both within and without the law in the sixteenth and seventeenth centuries. This renewed emphasis on racialized systems of bondage in the Spanish-held Indies complements other narratives of Spanish Empire that have emphasized the debates surrounding the conquest's legality, notably pursued by the canonical lawyers and theologians of the Schools of Salamanca, Alcalá de Henares, and San Gregorio de Valladolid, or the reiterated passage of Laws of the Indies, aimed at improving the lot of the Indians incorporated into the Spanish Empire as subjects. Thus, in the traditional historiography, 1542 figures largely as the date on which Indigenous enslavement was abolished and limits were placed on the leases of Indigenous labor held in encomienda. However, as Reséndez has shown, the legal abolishment of Indigenous slavery in 1542 did not end systems of Indigenous bondage. Indeed, "a new regime emerged in the 1540s and 1550s, a regime in which Indians were legally free but remained enslaved through slight reinterpretations, changes in nomenclature, and practices meant to get around the New Laws. All over the Americas, the other slavery took shape as Spaniards struggled to implement the laws" (Reséndez 2016, 126).

Here it may be useful to return to the Möbius strip of erasure and branding explored above. From branding bodies to erasing bondage, to what extent did abolishing Indigenous slavery render it invisible or illegible from the sixteenth century to the present? By returning to the chivalric fiction that Fernández de Oviedo sought to erase from his publication record later in life, I also bring into sharp relief his experience as slave brander of Indigenous persons in the Darién immediately prior to writing on the blank pages of his chivalric fiction. In other words, I do not claim to read new meanings into the *Claribalte* with reference to a biographical key; rather I contend the impossibility of reading the *Claribalte* as anything other than the first fiction written in the New World following a letrado's employ in a conquest as a slave brander.

Reading the *Claribalte* with the author's most immediate experience in the Darién in mind is an exercise in mapping lacunae, islands, if you will, to trace the chain of contingencies of New World fictions on the enslavement and subject making of legal performances such as the *Requerimiento* and the narratives constructed, documented, and certified by slave branders, such as Fernández de Oviedo, to justify the mark burned into the skin of so-called rebels against the Spanish imperial project. That Fernández de Oviedo himself sought to omit the *Claribalte* from his publication record while also erasing his official duties as a slave brander in the Darién from his personal history makes his stake in the legal fictions employed for the expropriation of the Indigenous peoples a key element for composing his contribution to the Latin American lettered tradition via branding but also erasure. It is a telling island in his own fictions of self-fashioning but one integral to the myth of the letrado whose stained hands, he protests, perhaps too much, are beholden only to the inkwell. How he produced and reproduced these fictions in the New World, both as a bureaucrat and as a writer of chivalric fiction, exemplifies the role played by lettered elites in both performing and erasing the bondage of the Indians and Africans in the Americas from the written record.

Rama's (1996) reading of the Latin American city as foundational to the region's lettered tradition is well known. The colonial city, according to Rama, was the result of a compulsion for order generated from the metropole; in addition to the reiterated checkerboard grids that distinguish Spanish American colonial cities, Rama cites the obsessive repetition of the word *order* in the contract between Pedrarias Dávila and Fernando II of Aragon. The writer charged with documenting the fulfillment of this order, which included an itinerary of preordained slave and subject making mapped out in the contract, was Fernández de Oviedo, who later received the position Official Chronicler to the Crown of the West Indies. He was also charged with certifying each performance of the *Requerimiento* and its outcome, in which he also had a financial stake as the slave brander of the expedition, detailed below. Following a brief excursion into the life and times of Fernández de Oviedo before writing the *Claribalte*, I show how the legal and mercantile tropes of the two scripts for this conquest, the *Requerimiento* and the capitulación, dovetail with the tropes of the chivalric and Byzantine novels.

Fernández de Oviedo, a hidalgo of Valdés, must have become familiar with the genre before he traveled to the New World, when by order of Fernando II of Aragon he served the duke of Calabria, also Fernando of Aragon (1488–1550), son of the deposed don Fadrique of Aragon (1452–1504). The library of don Fadrique, whose reputation as a bibliophile, predilection for the chivalric fiction genre, and literary patronage were later praised by Fernández de Oviedo himself in his *Batallas y quinquagenas*, would have provided a fertile ground for his encounter with the genre (Fernández de Oviedo 1989, 8). He later entered the service of Bishop don Diego Ramírez de Guzmán (?–1508) and in 1506 the Council of the Holy Inquisition. It seems that Ramírez advocated for his inclusion in Dávila's expedition, which set sail for Tierra Firme in 1513. As discussed in chapter 2, one of Dávila's instructions from Fernando II of Aragon was to rename the territory presently comprising Colombia and Panama, known as Tierra Firme (Mainland), Castilla de Oro (Golden Castile), in recognition of the precious metal found there in abundance.

Fernández de Oviedo joined the Dávila expedition as its undersecretary and was later named comptroller of the gold foundry (*veedor de la fundición de oro*); it was the latter official position that became a source of pride for Fernández de Oviedo, for he signs the *Claribalte* as the comptroller of the gold foundry in the book's prologue to his patron, Fernando of Aragon, Duke of Calabria. The author was less proud, however, of the other official duties he performed for Dávila's expedition, as undersecretary to Lope de Conchillos: "Fundición e mareación, e escribanía de minas, e del crimen e juzgado, e el oficio del hierro de los esclavos" (Smelting and navigation, notary of mines, and of crime and judgment, and brander of slaves) (Pérez de Tudela Bueso 1957, 418). As Pérez de Tudela Bueso noted, these other offices held by Fernández de Oviedo during Dávila's expedition would have been difficult if not impossible to fulfill with any empathy for the Indians (419). As slave brander, Fernández de Oviedo was responsible for scarring the skin of Indian slaves with the mark for *jure belli* (just war) for which he would receive 11 maravedís and his superior, Lope de Conchillos, would receive 56 maravedís per brand.[4] These fees, as stipulated in the capitulación between Fernando II and Dávila, were "lo que en sus conciencias dijeren e les paresciere" (determined by the consciences) (Morales Padrón 2008, 90) of the lord governor Dávila and Bishop Juan de Quevedo. In addition to officer of branding, Fernández de Oviedo held a partnership (*un asiento*) with Conchillos, who conceded half of all "rights" (*derechos*) produced by the secretary and subsecretary of the Dávila expedition; these derechos included their accounting for the raids (*cabalgadas y entradas*) equal to half a share of what would be distributed after each raid. For Fernández de Oviedo's participation as Conchillo's secretary in the raids, as official narrator and notary of these raids, he also received a salary of 40 pesos. Thus Fernández de Oviedo would have received remuneration from the stakes he held in the asiento for these raids while performing his official duties as scribe and notary of the raids, whose legality he was charged with chronicling and certifying, in addition to receiving a salary as a letrado. As analyzed in chapter 2, Fernando II of Aragon warned Dávila that the immediate incentives for slaveholding were greater than those for encomienda holding. I might add that for a letrado on such an expedition, the profit motive similarly favored slave branding and accounting for those brands in narratives that painted a picture of Indigenous monstrosity and insurgency, often conflated.

In mid-July 1514, Dávila arrived in the Darién with his wife and a cohort armed with weapons and documentation, most likely copied and notarized by Fernández de Oviedo himself, declaring that Dávila had been named the new governor of the territory, which would now be called Castilla de Oro. Vasco Núñez de Balboa, who had been serving as de facto governor, relinquished his house to the new governor and his family without a fight (Castillero 1957, 525). One year later, Dávila's terrible governance of Santa Marta and Castilla de Oro, which had left many in his cohort hungry and eager for raids on surrounding villages, would lead Fernández de Oviedo to leave the expedition prematurely in order to return to Spain, where we found him at the beginning of this chapter, seeking passage from Santo Domingo to Spain.

Perhaps that recent experience with hunger-driven raids in the Darién was refracted through the trope of shipwreck as described in chapter 59 of the *Claribalte*, which narrates the three-month-long penury of Félix and his sailors on Fire Island of Cape Verde; there, "no hallaron qué comer sino yervas y agua y no tantio déstas como quisieran" (they could not find ought to eat but grasses and water, and not enough as they would have liked) (Morales Padrón 2008, 90). Perhaps not. Unlike Félix, who convinces pirates to take him back to England, Fernández de Oviedo waited for a ship to return to Spain on the island of Hispaniola, where he wrote his *Claribalte*. In its quest for fictional diligence with the prerogatives of empire, Fernández de Oviedo's fiction avoids the opportunities offered by his narrative to explore the significance of traumatic events for the empire's agents as well as their beliefs about their moral and material superiority over the peoples they would subjugate in the name of Christian civilization. Unlike the later accounts of shipwrecks and other disasters written about figures such as Pedro Serrano, in Inca Garcilaso's *Comentarios reales*, or by Alvar Núñez Cabeza de Vaca, the civilization embodied by Félix is never placed in doubt. Whereas the recognition scenes in later narratives of shipwreck describe high-ranking Spaniards (a captain or a treasurer of an expedition) being mistaken for devils or natives by their fellow countrymen, Félix is immediately recognized by his admiral, Litardo, near the court in London (having tricked some pirates to gain a berth on their ship). As noted by Pinet (2010, 391), it is a classic scene of "ideological interpellation," for Litardo, the servant, recognizes his master—dressed as a sailor—because of his noble bearing. Disaster as affirmation of empire in the *Claribalte* may be the exception that proves the rule of shipwreck narrative in the Lusophone and Hispanic traditions when we account for the colonial context in which it was written. If, as Blackmore (2002) contends, shipwreck narratives constitute a form of "counter-historiography" in Iberian imperial cultures, Fernández de Oviedo's *Claribalte* bears more similarity to Daniel Defoe's *Robinson Crusoe* (1719) and its myth of the European male as the embodiment of civilization. As a fiction that hews to the topoi of Byzantine novels, it inverts Fernández de Oviedo's own experience with disaster in the Darién as governed by Dávila.[5]

The *Claribalte*'s main characters are armed bureaucrats, letrados. As Stephanie Merrim (1982) has argued, the novel's characters are motivated to document, certify, and declaim with little discursive room for action, though the novel does adhere, at least superficially, to the tropes of chivalric and Byzantine novels. And yet were it not for the biographical information supplied by the author, there is little in the chivalric fiction that points to Fernández de Oviedo's recent experience in the Darién because it hews so closely to the topoi of its genre. Nevertheless, the means behind the writing of the *Claribalte* is of interest precisely because it is produced by an author who chooses to write a chivalric fiction set in the so-called Old World while in transit from the New. A chivalric fiction written at the letrado's leisure, the *Claribalte* reveals the flipside of the business or busy-ness of conquest. This idleness that facilitates and even encourages writing after conquest may be juxtaposed to the perceived idleness of Indigenous subjects in the New World, which had led Isabel of Castile, years before the *Requerimiento* was redacted, to institute the encomienda (see Las Casas 1988, 1341–42). As he waited for a means to depart the New World, Fernández de Oviedo reproduced the scriptural economy of conquest while erasing its existence and his role in it for the purpose of his fiction. Or to repurpose Irving Leonard's (1992) now-classic concept, "books of the brave," the *Claribalte* participates in the kind of seamless translation, or the relation of metaphor, between Old World and New, and back again, that empire favors in the dominant narratives of its expansion.

Fernández de Oviedo's *Claribalte* adheres to chivalric commonplaces, but it reads like a fantasy of global dominion with the ugliness of enslavement and subjugation erased. The author affirms that he offers his readers but a translation of an original work that had been written in Tartar, which he found in the kingdom of Firolt. The hero's cupiditas and the impetus for his travels is set off by the invocation of Claribalte (whose name is "translated" as Félix), also a common trope of chivalric and Byzantine novels. Félix then wages a tour de force throughout Europe with the aid of a magic sword and necromancers. Set in a Christian era before the "discovery" of America, Claribalte's story never makes it as far as the New World, although he does suffer a shipwreck off the Fortunate Islands. Through strength of arms, magic, a fortuitous marriage, and the protagonist's illustrious lineage, Claribalte not only manages to kill the giants of the Isla Prieta (Black Island) but also secures the crown of emperor of Constantinople and the seat of supreme pontiff of the Christian world in Rome.

Thus the author's synthesis of religious and temporal powers in the figure of the chivalric knight reconciles the stark contrasts between interests held by the church, state, and investors in individual enterprises. By constructing the narrative climax around the fusion of religious and temporal powers, Fernández de Oviedo's *Claribalte* depicts a unified terrain of European dominion. This depiction of continental unity, however, could not be further from the reality on the ground: an emerging schism in the church with the Reformation and failed attempts to reconcile differences between the Orthodox and Roman Churches. Yet Claribalte's double throne accomplishes in chivalric fiction the pretensions to temporal and religious dominance outlined by the conquest's general partners in the *Requerimiento*, which presented the universal authority of a church whose seat was in Rome but *could be moved anywhere* and a seamless continuity between apostolic Christianity and the imperial pretensions of the Spanish crown in the sixteenth century. As presented by the *Requerimiento*, the church's universal dominion had allowed its supreme pontiff, Alexander VI, to "give" part of the world west of the Cape Verde and Azores Islands to Isabel of Castile and Fernando II of Aragon and their descendants. However, both the papal bull *Inter caetera* (May 4, 1493) and the Treaty of Tordesillas (June 7, 1494) were promulgated *after the fact* of capital investment and returns on investment in Portuguese and Spanish journeys of exploration west and east of the Iberian Peninsula. The contradictions of this metaleptic narrative were not challenged by Fernández de Oviedo but did receive detailed exploration by Las Casas in the "digressions" to his larger narrative of the conquest of the (West) Indies and traced in the section below. Fernández de Oviedo's chivalric fiction does not iron out these temporal and causal wrinkles in his first work as an author; rather he indulges in the same kind of reconciliation that must have been difficult to accomplish as the letrado on the Dávila expedition.

Though Fernández de Oviedo did not set *Claribalte* in the New World, as noted above, he does include a shipwreck off the Fortunate Isles, the name given to the Canary Islands, or what Gruzinski (2014) has called the laboratory for globalism and Wallace (2004) the *prefigure* for the conquest of America. For now, Fernández de Oviedo's incursion into adventure writing ideology serves as a reminder of the various fictions at play in conquests, including the legal ones, prior to 1492. More than a century before the Spanish conquest of the Canary Islands, dated to 1488, Luis de la Cerda received the title *Caballero de la fortuna* from the pope in Avignon in the late fourteenth century.[6] Like the trajectory of the eponymous knight-errant in the *Claribalte*, who, like don Luis de la Cerda, also assumes the epithet "Caballero de la fortuna," the narrative of conquista blurs causes and effects. Yet there is some historical truth, in the Foucauldian sense, to *Claribalte*'s chivalric fictions in terms of how claims to conquest were made: the fact of religious and temporal dominion rewards the impresario after a difficult and arduous journey, not before.

Having made the trip to and from Spain several times, Fernández de Oviedo channeled his imperialist pursuits as self-appointed royal historiographer in 1532. The *Historia general y natural de las Indias* (1535, Primera Parte) and its *Sumario* (1526) exploit the chivalric quest for its narrative horizon: the cornucopia or grail. By offering his readers a treasure trove, that is, a *thesaurus*, of exotic commodities and moralizing anecdotes, the royal historiographer pursued the temporal and religious dominion visualized at the close of the *Claribalte*. Rather than follow the trajectory of one knight-errant, Fernández de Oviedo offered up his pen to document all the wealth and exploits at the service of the conquest. An ardent defender of the Spanish crown's rights to the Indies, Fernández de Oviedo enjoined his fellow conquistadors in their pursuit of profits through violence. Though he did condemn the use of indiscriminate violence against the native peoples of La Florida by the failed expeditions of Pánfilo de Narváez (1478–1528) and Hernando de Soto (ca. 1496–1542), his condemnation of *indiscriminate* violence, as Rabasa (2000) contends, begs the question of what exactly constitutes a *discriminating* use of violence. In effect, Fernández de Oviedo urged his peers to pursue a cost-benefit analysis of their violence.

Legitimacy and efficiency dovetailed in Fernández de Oviedo's program for a successful empire. Knowledge of the peoples in the process of becoming subjects of the Spanish Empire complemented the textual cornucopia of riches he offered his readers. Within Fernández de Oviedo's ethos of Spanish and Christian dominion over the West Indies, the bounty of knowledge turned epic narrative on its head. Rather than define the horizon of desire for the quest, the treasure trove—plus the thesaurus—becomes the means and end of capital-funded ventures. Like the discomfiting oxymoron of the capital-begetting process, reconnaissance for the sake of creating more knowledge to inform even more quests obeys its own self-replicating dynamic: the preferred trope of scalability that functions contiguously but projects itself mimetically. For the proud comptroller of the gold foundry, it was only fitting that crown officials gave him a berth on a treasure galleon on his first trip back to the Iberian Peninsula from Santo Domingo.

ISLANDS OF RABBITS AND WOOD

Of Fortune and Wood, the Cannibal Isles, the Indies: these are strange names for places, and their stories are difficult to tell, however briefly. When rendering destruction via summation, *en resumidas cuentas*, as Bartolomé de Las Casas did in the account he published in 1552, he constantly alluded to people and events that he could or would not name but that could be found instead in the *Historia de las Indias*. While the *Brevísima relación de la destruición de las India's* narrative of destruction begins seemingly at the beginning, with the expedition led by Christopher Columbus in 1492, it is in the *Historia* where Las Casas seems to be at a loss as to how to tell the tale, and the digressions multiply even as the stakes had never seemed higher.

Unlike the *Brevísima relación*, which, on the surface, focuses on the Spanish Conquest chronologically and geographically, the *Historia de las Indias* does not hew to the original outline for the narrative of conquest. As I mentioned earlier in this chapter, when Las Casas began writing the *Historia de las Indias*, he originally planned to dedicate one volume to each decade of the Spanish Conquest from 1492 to his present. As he wrote over the course of thirty-four years, however, the question of when and where to begin his narrative shaped the emerging story line, and another historical project emerged: the comparatist *Apologética historia sumaria*. Where the *Apologética* provided greater depth about the culture and peoples of the Americas before the conquest, the *Historia de las Indias* continues to expand its geographic reach to include "the Indies" east of the Tordesillas treaty line. It is also in the *Historia de las Indias* that Las Casas revisited his early advocacy in the *Memorial de remedios para las Indias* (1518) for the increased importation of African slaves. His examination of the conquests in Africa before and after 1492 would lead him to write in the third book of the *Historia de las Indias*:

Este aviso [*Memorial* de 1518] de que se diese licencia para traer esclavos negros a estas tierras dio primero el clérigo Casas, no advirtiendo la injusticia con que los portogueses [*sic*] los toman y hacen esclavos; él cual, después de que cayó en ello, no lo diera por cuanto había en el mundo porque [desde entonces] siempre los tuvo por injusta y tiránicamente hechos esclavos; porque la misma razón es dellos que de los indios. (Las Casas 1994, 2191)

———

[The recommendation given [in the *Memorial* of 1518] to award licenses in order to import black slaves was first given by the cleric [Las] Casas, who did not know that the Portuguese captured and enslaved them unjustly; after learning of this [injustice], he would not give the same recommendation for all the world because [ever since] he has held that they are unjustly and tyrannically enslaved; because it is the same with them as with the Indians.]

In a mode similar to his interpellation of Columbus, in hindsight Las Casas questions the role played by his ignorance in the enslavement of black Africans. This is a recurring theme in the *Historia de las Indias*, for Las Casas had conceptualized both slaveries in connection to each other through his digression from the story of Columbus to the Portuguese origins of African slavery in chapters 22 and 27 of the first book (Las Casas 1994, 459–93). By condemning the bondage of black

Africans—and himself in the third person—Las Casas was effectively criticizing the fourteenth- and fifteenth-century enterprises in Africa, which had served to develop the Iberian styles of monarchic venture capitalism for global scalability.

In one of the *Historia*'s first of various digressions, Las Casas maps the trajectory of various capital and military campaigns in the Canary (or Fortunate) Islands, Madeira Islands, West Africa, and Goa, highlighting the corporate and multinational state of profitable violence. In noting this apparent digression, let us return again to Dipesh Chakrabarty's contention, in *Habitations of Modernity* (2002, 31), that origins—especially violent origins that give way to modernization processes—lure the intellectual into redacting their narration, especially when the intellectual feels implicated in the legal conservation, through state powers, of that foundational violence.

What are we to make of Las Casas repeatedly indicating one or another event as the origin of systematic oppression, up to his very present? He shuffles the order of events. For the emplotment of conquest in the *Historia de las Indias*, first he tells of Columbus, then of the Canary Islands, then of Madeira (Porto Santo and others), West Africa, Goa in the East Indies, and back again. Why weave such a tangled web? For this section on the Portuguese conquests in Africa (Guinea), including Columbus's involvement in Madeira and in the enslavement *feitoria*, or trading post, in Mina, Las Casas leans heavily on *Ásia* (1552) by João de Barros (1496–1570), who published his celebration of Portuguese imperial enterprise in the same year that the *Brevísima relación* was published in Seville, without the customary license, valuation of pages, or censure for publication in this time period. This reliance on Barros's material, combined with his disputation of the Portuguese author's celebration of Portuguese empire, underscores the global and transimperial turn Las Casas's approach took following the *Brevísima relación*, whose subsequent reception, translation, and propagation in Protestant communities led to the creation of the black legend, that is, the narrative that attributes, among other things, an exceptional nature to Spanish Catholic violence in its overseas colonies.[7] The *Brevísima relación* itself belies such an interpretation, since it also dedicates one chapter to the Fuggers (*Fuqueros* in Spanish), the German bankers turned conquistadors, in Venezuela.[8] However, one of the main achievements of the *Historia*'s sections written after 1552 is to widen the historical lens to include Portuguese enterprises. One historical anecdote about Bartolomé Perestrello's doe rabbit on the island of Madeira serves as an allegory that, in the sense originally explored by Victor Shklovsky in his seminal essay "Art as Device" (1917), defamiliarizes common tropes of *natura* and *natural* in conquest narratives.

Just an island within the larger narrative archipelago of the conquest writ large, the anecdote about Bartolomé Perestrello's pregnant doe rabbit in Las Casas's *Historia de las Indias* offers a visualization of conquest that blurs the binaries by which the conquest operated discursively. In this anecdote, conquistadors of several nationalities were involved, and no *naturales* were harmed, for Porto Santo in the Madeira archipelago was a deserted island. Yet this is the island where Columbus would prove his mettle, where he would father his firstborn son, where the landscape would first be destroyed by an invasive species (the hare) and almost immediately thereafter by capitalism (in the form of logging and shipbuilding), a cash-crop economy (sugarcane), and becoming a stepping-stone in the increased commodification of human beings for their labor and for their souls. Moreover, the rabbit as sign offers its own warnings on univalent narratives. For what the sailors on Bartolomé Perestrello's ship first interpreted as a good omen—the many births from a pregnant rabbit transported over the Atlantic Ocean in a hutch—in hindsight became a portent of rampant destruction.

In the Spanish of the long sixteenth century, *naturaleza* could refer both to a subject's community of belonging—synonymous with *patria*—and to nature, the latter understood tautologically as those living beings defined as *not* human by some humans. In the context of an expanding empire through overseas *empresas* (enterprises in the wider sense), the symbolic and material economies of the natural and belonging(s) rely on classifications and definitions that seem both transparent and contradictory. Taking into account Shell's observations in *Islandology* (2014) that islands have always served as schematic shorthand for classifications and categories (as in Venn diagrams) and also for indigeneity, an anecdote about a plague of rabbits on the Madeira Islands in the early fifteenth century blurs the limits between Portuguese and Spanish, animal and human, native and nature (*natural y naturaleza*), business and law, constitutive and constituted violence, between *natura* and *contra natura*. For the editor of Columbus's *Diario* and author of the *Historia de las Indias*, the telling of *the* history of the Indies also involved another kind of blurring, between the story of the man and the lands that he supposedly discovered but, Las Casas was one of the first to recognize, had their own stories to tell.

Whereas Las Casas's *Historia de las Indias* begins with divine providence, the first entries of Columbus's *Diario*, as edited by Las Casas, open with disaster. But three days at sea, the *Pinta*'s rudder had jumped; her hull was leaking. In his ship's log for August 6, Columbus wondered whether the *Pinta*'s rudder had come loose from its fastenings or if, perhaps, one Gómez Rascón and one Cristóbal Quintero, the owners but also the troublemakers among the *Pinta*'s crew, had loosened it (Columbus and Las Casas 1989, 22).[9] Columbus and company had set sail from Palos, a small portside town in southern Spain, just three days earlier. Palos was not a natural first choice. That would have been Cádiz or Seville, for these cities boasted larger ports, more ships, and more seasoned crews. But Columbus made do, for in early August 1492, the other, larger harbors were teeming with people and ships headed south and *east*. After centuries, the Jews of Sefarad—known as Spain to the Christians—were being forced from their native Spain into exile. In the early days of January 1492, Isabel of Castile and Fernando II of Aragon had defeated Granada, the last Muslim-held kingdom in Europe. On March 31, the Catholic monarchs signed the Edict of Expulsion, ordering the Jews of Spain to convert or depart in three months' time or face death. The edict referenced the Spanish Inquisition's fears of contagion and contact between practicing Jews and the new Christians and argued that it was not enough to have quarantined the Jews in ghettos, as they had been ordered to do beginning in 1480. "Así los naturales dellos como los non naturales" (Both natives and non-natives), the monarchs contended, the Jews of Spain no longer belonged because "pervierten el buen é honesto bevir de las ciudades e villas, é por contagio que puede dañar a los otros" (they pervert good and honest life in cities and towns, and by contagion can hurt others) (Fita 1887, 516–17). So, natives or not, the Jews of Sefarad were treated as invaders, at about the same time that the newly constituted body politic known as Spain set its sights south and west for its own overseas invasions.

Thus, as Menocal (1994, 5) has noted, in early August 1492, "the voyages of exile and the voyage of discovery begin at the same hour, in the same place," though Columbus himself barely mentions the forced exile of the Jews in his *Diario*. For if anti-Semitism has always proven itself useful for providing a "general framework of the development of the nation-state," as Hannah Arendt (1966, 17) argued in *The Origins of Totalitarianism*, such frameworks go sight unseen more often than not, especially when they depend on the banishment of the other from our midst, on their invisibility. Guided by Columbus, we train our gaze instead on those three ships departing from Palos, for in the prologue to his *Diario*, addressed to Fernando and Isabel, he chooses to underscore instead that the *Niña*, the *Pinta*, and the *Santa María* were very well suited for transoceanic travel (*muy aptos para semejante fecho*), even though his subsequent diary entries disavow this rosy picture of the ships almost immediately (Columbus and Las Casas 1989, 18–19). As Columbus's canonical biographer, Samuel Eliot Morison (1942), and Menocal (1994) later observed, Columbus turns his gaze all too quickly from the scenes of despair he left behind on the docks of southern Spain in early August, even though the exile of the Jews had compromised the material conditions of his expedition. And it is all too easy for us to turn our gaze with him, to the wake of those three ships.

It bears repeating: after just three days at sea the *Pinta* was both leaking and rudderless. But Columbus and his men would find a safe harbor in Lanzarote and Tenerife on the Canary Islands. It was a *safe* harbor in that it was a *colonized* harbor. Conquered in 1488 by the forces of Isabel and Fernando, after two centuries of disputed claims among European trading companies and monarchs, the Canary Islands offered a stepping-stone to the lands and peoples that lay to the southwest.[10] There, the *Pinta* was recaulked, her rudder replaced, and her sail rerigged, though Columbus would have preferred to exchange her for another, sounder vessel, he confesses in the *Diario* (Columbus and Las Casas 1989, 24–25). All this was done in just under a month, which gave Columbus some time to recall the origins of his voyage while he and his men waited.

Columbus wrote but one diary entry for that entire month when he could have been at sea. Beyond recalling the arrival of a man from the Madeira Islands to Portugal some eight years earlier, Columbus merely records that this man had requested a ship from the king of

Portugal, John II (r. 1455–95), to travel west to the lands he claimed to have seen on the horizon each year from Porto Santo. According to Columbus, the man from Madeira was not the only one to make such a claim. Similarly, the new residents on the Azores were known to declare that on a clear day lands could be discerned on the western horizon.

Similar to those Portuguese transplants to the Azores, Columbus was himself a man from Madeira through his marriage into the Perestrellos, a Genovese seafaring family based out of Lisbon. While his own origins remain obscure, we know this much about Columbus: were it not for his time on Madeira, Columbus most likely would *not* have set sail from southern Spain in August 1492. Were it not for the exile of the Jews from Spain, Columbus would *not* have set sail from the lesser port town of Palos. In his edition of the *Diario* and the *Historia de las Indias*, Las Casas asks us to make sense of such contingencies as we turn our gaze southwestward and try to connect these stories that hinge on cultural and political economies of naturaleza: of contagion and contact between those *not* belonging.

To recapitulate, Las Casas will begin at the beginning in the *Historia de las Indias*, he tells his readers over and over again. If he begins, as I have, in 1492 to explain how the West Indies became known as "the Indies" via a providential error made by Columbus, by chapter 4 of book I of his *Historia* Las Casas finds himself reaching further back, to tell the story of how Christopher Columbus came to be Columbus through marriage. After all, it was Columbus's mother-in-law, the wife of Bartolomé Perestrello (1395–1457), who recognized Columbus's interest in seafaring, Las Casas argues. The wife of Perestrello remains unnamed, but she was the one who shared with her son-in-law the instruments, narratives, and maps relevant to the conquest of the west coast of Africa, which was known as Guinea at the time. And yet by introducing the Perestrellos into the story of Columbus, which is also the story of the New World, Las Casas must answer the question how it was that the Perestrellos came to live on the Madeira Islands. After all, they are of Madeira but not *native* to it. This realization begets another narrative incursion, further back in the past, and it involves, of all things, a very pregnant rabbit borrowed from Barros. But I'm getting ahead of myself, and of Las Casas.

After the mother-in-law anecdote, Las Casas offers a guidepost to his readers, noting that the history of the discovery of Madeira will be told starting in chapter 17. He also references the *Ásia* of João de Barros as one of his sources. Returning to his biography of Columbus, Las Casas then notes how from the stepping-stone of Porto Santo in Madeira the future admiral of Spain joined other Portuguese expeditions to Mina, where he learned the slave trade. In Madeira, where the Perestrellos had left Columbus a hacienda and other properties, Columbus became a father for the first time, as it is there where Diego Colón (1479–1526), his future heir to the Spanish title Admiral of the Ocean Sea, was born.

As Las Casas tells it, Diego Colón became admiral to the Spanish crown in 1519 in Barcelona, the same year and the same city where Charles I arrived in Spain for the first time. In a thinly veiled comparison between the new Spanish admiral–Genovese merchant and the new Spanish king–Christian emperor, who were both born outside of the Iberian Peninsula, it was also in Barcelona where Charles I of Spain received the news of his imperial election and new title (Charles V, Holy Roman Emperor), Las Casas observes. Las Casas goes on to emphasize the importance of Porto Santo in the Madeira archipelago as the place where Christopher Columbus "engendró al dicho su primogénito heredero don Diego Colón" (conceived his firstborn son and heir, Diego Colón) (Las Casas 1988, 3, 368). This Porto Santo, or Holy Port, is also where Columbus would be inspired to look southwest for another route to the Indies, like the other new, restless transplants in Madeira and the Azores who were also setting their sights south and west once more.

As a stepping-stone, Porto Santo is both fertile ground for empire and, as Las Casas tells it, a wasteland. It is but one of the repeating islands of destruction, if I may borrow the phrase from Antonio Benítez-Rojo (1996) and extend his gaze east and further back in the past, to the longer archipelago of colonial capitalism, the seemingly inexorable replication of wastelands of the long sixteenth century, the phrase coined by Braudel and recalled by Arrighi to describe the period defined by the expeditions fueled by Iberian monarchic capitalism (ca. 1450–1650). By Las Casas's reckoning, Porto Santo was thus fertile soil, figuratively but in a sense quite literally, for the imperial expansion of both Iberian powers, even as its soil became depleted by colonial settlement and the cultivation of sugarcane. It became a launching site for expeditions east and west of the islands of Madeira, first exploited for their wood—hence the metonymy—and, immediately thereafter, the first among a chain of sugar-producing islands. Thus the Canary and Madeira Islands are first in a chain of capitalism's re/production of our current world ecology, to reference Moore's (2015) term. With the promise to his readers that he will return to the west coast of Guinea and the mid-Atlantic islands in later chapters, Las Casas closes his first digression on Madeira.

It is not until chapters 17 through 22 of the second book of the *Historia de las Indias* that Las Casas maps a tangled web of financial, legal, and social partnerships that illustrate how the 1494 Treaty of Tordesillas might have etched a line that was a distinction without a difference between the Portuguese and Spanish Empires. Porto Santo would be annexed to Portugal as one of the Madeira Islands, but, as Las Casas argues, the conquests of places named for Fortune (the Canary, or Fortunate, Islands) and Wood are, to recall Lola Rodríguez de Tío's imagery for Puerto Rico and Cuba, in a different century, a different time, "the two wings of the same bird," in this case, an *azor*, a bird of prey, for which the Azores themselves were named. In other words, Las Casas explores how the sense of belonging was constructed—the Canary Islands belong to Spain, the Azores to Portugal—by reconstructing the respective invasions of these islands by foreign powers.

So Las Casas begins the five-chapter foray into the mid-Atlantic and West Africa with the Canary Islands and the Azores. It opens with a detailed description of the *financial* schemes employed by various empresarios—French, Castilian, Aragonese, and Portuguese—and how various European monarchs did (or did not) back the colonial aspirations of their respective subjects. As Las Casas demonstrates, the original charters for the conquests of these islands in the early fifteenth century did *not* delineate precisely between private ownership of enterprises in conquest, the nationalities of the entrepreneurs, and the monarchs who had chartered them. This seems counterintuitive because we might think that the movement toward globalism would begin with questions of "national" jurisdiction and dominion to which territorial claims would be appended by expanding imperial and capitalist states.[11] However, in the late fourteenth and early fifteenth centuries, the primacy of the trading companies made jurisdiction secondary to private gains or, at least, into a point of contention. This was the case of Maciot Betancourt, a Frenchman, who inherited the island of Gomera from his uncle, Jean Betancourt, whose expedition had been made up of Frenchmen and Castilians under a charter from King John II of Castile (r. 1406–54) or the queen regent, Doña Catalina (r. 1406–18). However, Jean Betancourt's funds soon ran out while fighting the Guanches, the native Canary Islanders, and when Castile was not forthcoming with more capital, Jean and Maciot Betancourt sold their conquest to Prince Henry the Navigator (1394–1460) of Portugal and moved to Madeira, which in 1419 had been discovered by Bartolomé Perestrello, the Genovese-Portuguese merchant and future father-in-law of Christopher Columbus. Soon thereafter, Prince Henry of Portugal sent an armada to back up his capital investment.

Las Casas reproduces the ensuing confusion generated by these private enterprises in imperial violence by juxtaposing the letters exchanged between John II of Castile and Alfonso V of Portugal (r. 1438–81), in which at various points the monarchs—also family and rival claimants to the crown of Castile on the peninsula—laid claims to the island either via synecdoche, because one of "their" subjects had performed the conquest in question, or via metonymy, because the monarch's capital had purchased dominion. My point and, I believe Las Casas's point, pace W. K. Wimsatt and M. C. Beardsley (1946), is not to promote the claim of one monarch over another to the Canary Islands, for example, though Las Casas ostensibly does just that at times, but to note that experiments in profitable violence in the fifteenth century did *not* lead, inexorably, to the regulated practices of conquest in the sixteenth century. Though it would become commonplace for Portuguese navigators, for example, to lead ventures for Spanish monarchs (think of Ferdinand Magellan in the Pacific Ocean) or for these ventures to be multinational beyond the Iberian Peninsula in their makeup, including Greeks, Frenchmen, Germans, and Italians, the common understanding by the sixteenth century was that the chartering monarchy laid claim to the conquered (or to-be-conquered territories), not the monarch of the subject doing the conquering. By extension, the writing practices of conquistadors would similarly follow the norms laid out by the laws in the country of charter, yet these laws arose out of the process of trial and error in which the Iberian monarchs had engaged in the fourteenth and fifteenth centuries. By exposing in minute detail the genealogy of commonly held practices of conquest, Las Casas, in effect, disentangles the habitus of the sixteenth century, so imbricated with that of conquest but also expulsion.

It is with the chapters on the conquest of the Canary Islands (chapters 17–20 of the first book of the *Historia de las Indias*) that Las Casas can explore the arbitrary nature of imperial dominion and jurisdiction, for by the close of chapter 20, Las Casas has given his readers the entangled and confusing history of the financing and political schemes behind the Canary, Madeiras, and Azores Islands' conquest by Europeans of various naturalezas (communities of belonging), with the islands' ensuing depredation.

And so it is in chapter 21 that Las Casas explores the livelihood of the native Canary Islanders, an alternative habitus to the life and times of conquest. The ingrained practice of profitable violence is thus juxtaposed to the Canary Islanders' natural, and "also native" (*también natural*), ways of relating to their land and to one another. In the next and final chapter of this section, Las Casas returns to the Perestrello-Columbus connection as promised in the first book of the *Historia de las Indias* and, thus, to Madeira. He chooses to close his narrative incursion into the islands off the western coast of Africa with an anecdote about a plague of rabbits, an invasive species brought by Perestrello on his ship in 1419.

While lack of space prevents me from analyzing the discursive complexities created by Las Casas's essentializing strategies in this chapter, I would like to underscore what he achieves with this play on the double meaning of *natural* in Spanish and its possibilities for contesting the logic of *contra natura* and the use of this category in conquest.[12] If we recall that observation and recording of practices considered contra natura by naturales in their naturaleza constituted grounds for their enslavement and death, then the attempt to blur these categories, at least in part, deserves our critical attention. Much as Horswell (2005) and Rabasa (1993) did for the *Apologética* by Las Casas, which we recall sprang from his *Historia de las Indias* project, I show how Las Casas queers the classification scheme performed and produced by coloniality while at the same time remaining—at times—complicit with it. A pregnant rabbit on an island, a mere anecdote in the three-tome work that is the *Historia de las Indias*, provides insight into the thought, but also the activism of, Bartolomé de Las Casas, on the limits of the human and its slippery relation to the category "nature."

Before Perestrello and his men reached the shores of the island that he would name Porto Santo, of the islands later named for their prized commodity, Wood (Madeira), they witnessed a doe rabbit giving birth on the ship.

> Entre otras cosas que llevó el Bartolomé Perestrello para comenzar su población, fue una coneja hembra preñada, en una jaula. La cual parió por la mar, de cuyo parto todos los portogueses fueron muy regocijados teniéndolo por buen pronóstico de que todas las *cosas* que llevaban habían bien de multiplicar, pues aun en el camino comenzaban ver *fructo* dellas. Este *fructo* fue después tanto y tan importuno que se les tornéo en gran enojo y en cuasi desesperación de que no sucedería cosa buena de su nueva población; porque fueron tantos los Conejos que de la negra, una y sola coneja se multiplicaron, que ninguna cosa sembraban y plantaban que todo no lo comían o destruían. Esta multiplicación fue tanta y en tan excesiva numerosa cantidad por espacio de dos años que, teniéndola (como lo era) por pestilencial e irremediable plaga, comenzaron todos a aborrecer la vida que allí tenían. Viendo que ningún *fructo* podían sacar de sus munchos trabajos casi todos estuvieron por se tornar a Portogal, lo que al fin hizo el dicho Bartolomé Perestrello, quedándose los otros para más probar, porque la divina providencia tenia determinado por medio dellos descubrir otra isla donde su santo nombre invocar y ser alabado. (Las Casas 1988, 463; emphases mine)

> [Among other things that Bartolomé Perestrello took to start his settlement [*población*] was a pregnant doe rabbit, in a cage. She gave birth at sea, from whose birth all the Portuguese rejoiced, taking it as a good sign that all the *things* they carried would multiply, because even en route they had begun to see their *fruits*. This *fruit* was later so prolific and so annoying that they got angry and

almost despaired that no good thing would come of this new settlement; because there were so many rabbits [born] from that black rabbit that multiplied, there was nothing [the Portuguese] did not sow or plant that they [the rabbits] did not eat or chew. Over two years, this multiplication was so great and excessive that, concluding it to be a pestilential and irremediable plague (as it was), they all began to abhor their life there. Seeing that their many works could not bear any *fruit*, almost all of them were about to return to Portugal, to which Bartolomé Perestrello finally did, and the others remained to try some more, because divine providence had determined through them to discover another island where his Holy name would be invoked and praised.]

He is referring to Madeira (Wood), "que después y agora tanto fue y es provechosa y nombrada" (which later became, as it is now, so beneficial and is thus named) (Las Casas 1988, 463). Paradoxically, the island of Porto Santo is a wasteland.

In the passage quoted above, Las Casas mentions "fructo" no fewer than three times in the paragraph dedicated to Perestrello's rabbits, as in the rabbit bore fruit, but the rabbit is also just another thing (*cosa*) that bore fruit in the ship's hold. When *fructo* appears for the first time in the passage, the rabbit's progeny is interpreted as a sign of good fortune. For the traders, soldiers, and merchants on the caravel, the rabbit's fertility signifies their future wealth; just as the rabbits multiply, so too will the other "things" in the ship's cargo. Yet what had been interpreted as a sign of good fortune becomes a plague on the island, so life there becomes unbearable. Life as commodity and reproductive life as commodity replicate destruction—not the "creative destruction" that Joseph Schumpeter (1942, 82) would famously define as "the essential fact about capitalism" but, it bears repeating, *destructive destruction*.

The passage's emphasis on things bearing fruit, and the language that treats the rabbit as if it were a thing, is suggestive of Thomist critique of usury, discussed in chapter 1. As a reminder, Aquinas, following Aristotle, famously noted that money is "sterile," unlike a tree, and thus could not bear fruit. Thus money multiplying (i.e., generating interest) was against nature (contra natura). In this passage, both "things" (cosas) and rabbits multiply first on a boat and later on an island. Rabbits and things are interchangeable for their potential to "bear fruit": the ship and the island seem equally exchangeable as places for production and reproduction. At the same time, the "natural" way in which the rabbit "bears fruit" is also contra natura, in the sense that the rabbit was not native (not a natural) of the island. In this way, in a few lines the anecdote destabilizes the seemingly transparent opposition between natura and contra natura and yet simultaneously interrogates the way in which this form of colonizing capitalism blurs the distinction between things and beings.

Las Casas literally adds color to the passage by underscoring that the doe rabbit is black (*negra*); Barros gives no color for the animal in the original. In Las Casas's rendering, both "things" and the "black" animal reproduce on a Portuguese ship heading toward Africa and wreak destruction. As an allegory for chattel slavery, it almost seems too transparent. If I may rephrase the opening play on supplement and being from Derrida's *Beast and the Sovereign* (2009), the rabbit is/and (*est/et*) the sovereign and/is (*et/est*) also the capitalist. Or, as Christian Parenti (2016, 182) contends, the state "does not *have* a relationship to nature—it *is* a relationship with nature" (original emphasis). The rabbit multiplies on the Portuguese ship of state with ramifications for Iberian modus operandi globally.

The passage about the plague of rabbits is a translation of Barros's *Ásia* (I.ii) with some minor, yet salient, differences that point to the direction of Las Casas's critique of the binary structures of conquest. In the original passage, the rabbit's labor on board seems to have a magical effect on the seeds the ship is also carrying.

> Entre as quais numa coelha que Bartolomeu Perestrello levava prenhe metida numa gaiola que pelo mar acertou de parir, de que todos houveram muito prazer: tiveram por bom prognostico, pois já pelo caminho começavam dar fructo as sementes que levavam, aquela coelha lhe[s] dava esperança da multiplicação da coelha os nem enganou, mas foi com mais pesar que prazer de todos: porque chegados a ilha solta a coelha com seu fructo, em breve tempo multiplicou em tanta maneira, que nem semeavam ao plantavam cousa que logo nem fosse roída. O que foi em tanto crescimento por

espaço de dos anos que ali estiveram, que quase importunados daquela praga, começou de aborrecer a todos o trabalho e modo de vida que ali tinham: donde Bartolomeu Perestrello determinou de se vir pera o reino, ou per qualquer outra necessidade que pera isto teve. (Las Casas 1988, 16)

[Among these things was a doe rabbit, which Bartholomé Perestrello brought in a hutch, and gave birth at sea, which pleased everyone; they took it as a good omen, for already the seeds they brought with them had started sprouting; that rabbit gave them hope by her multiplying, but she deceived them, for she ended up giving them more pain than pleasure; because once they had arrived on the island and received the rabbit and her fruit, they shortly multiplied so quickly that they [the Portuguese] could not plant a thing that would not end up chewed up. And their growth was such during the two years of their stay, and they were so annoyed by this plague, that their life and labor there became abhorrent to them; so that Bartholomé Perestrello decided to return to the kingdom [of Portugal], or for another reason that he had at the time.]

The seeds bear fruit as expected in the "natural" order. In the translation by Las Casas, the gloss of things for seeds would seem to hint at something out of sorts, for "things" could be construed so generally to include re/productive practices of an unnatural order. Here Las Casas plays with the double meaning of *natural* in Spanish, with the imagery of the rabbit bearing fruit, repeatedly, and her progeny multiplying, the logic of capitalist production *from* nature's vital forces exposed: re/producing destruction not only on Madeira, but beyond. Natural but not native, the destructive and reproductive capacity of Perestrello's pregnant rabbit allegorizes the capitalism of conquest as a double internality of humanity in nature. But all of humanity? In part? Can anyone or anything remain without the logic of primitive accumulation?

These lessons about the relations between humans, nature, capital, and the state learned by Columbus with the Portuguese and the Genovese on Madeira and thereabouts he later applied to the Indies in this hemisphere. In the *Diario* entry for Sunday, November 25, 1492, Co-

lumbus foreshadows the figure of the manufacturer who claims ownership of a waterfall, later explored by Karl Marx in the third volume
of *Capital*. In this passage, Marx (1993, 3:782–84) contends that a monopoly over what he calls the "motive force of power" of nature leads
to a surplus of profit for the capitalist, which is independent of labor.
In the *Diario*, Columbus offers an insight into the consciousness of the
manufacturer and his colonizing vision for the waterfall, which sub-
sumes natives as another form of nature to be exploited for profit. He
classifies the landscape in terms of its potential to generate a cluster of
economic activity (such as gold panning, logging, shipbuilding) and
anticipates violence against the land's Indigenous people (naturales).
He begins by giving a vision of scale for the land's topography rendered
in a unit of measurement that, I fear, can become familiar or even
natural to readers of colonial Latin American texts: a cape that defines
a deep, natural harbor—he tells his readers, the Catholic monarchs of
Spain—is at a distance of "two crossbow shots" (*dos tiros de ballesta*).
Such a measurement only makes sense within a logic of capitalist
violence that foresees the need to "defend" the stockade of this land
from its native inhabitants.

> Vio venir un grande arroyo de muy linda agua que deçendia de una
> montaña abaxo y hazia gran ruydo. Fue al rio y vio en el unas pie-
> dras relucir con unas manchas en ellas de color de oro: y acordose
> que en el rio tejo que al pie del junto a la mar se halla oro y pare-
> cionle que cierto devia de tener oro/. Y mando coger çiertas de
> aquellas piedras. Para llevar a los reyes. Estando asi dan bozes los
> moços grumetes diciendo que vian pinales. Miro por la sierra y vi-
> dolos tan grandes y tan maravillosos que ~~no le~~ [?] podía encareçer
> su altura y derechura como husos gordos y delgados donde cogno-
> scio que se podían hazer navios e infinita tablazón y masteles para
> las mayores naos dspaña. Vido robles y madroños y un buen rio y
> aparejo para hazer sierras de agua . . . ido por la playa munchas
> otras piedras de color de hierro: y otras que dezian algunos que
> eran de minas de plata todas las quales trae el rio. alli cojo una en-
> tena y mastel para la mezana dela carabela niña. Llego a la boca del
> rio y entro en una cala al pie de aquel cabo de la parte del sueste
> muy honda y grande en que cabrian cient naos sin alguna amarra

ni anclas. Y el puerto ~~tal~~ que los ojos otro tal nunca vieron. . . . Y muestra de aver resçibido de verlo y mayormente los pinos inextimable alegría y gozo. [al margen: porque se podía hazer allí quantos navios desearen trayendo los adereços si no fuera madera y pez que allí se haría harta] y siempre en lo que hasta alli avia descubierto yva de bien en mejor./ ansi en las tierras y arboledas y yervas y frutos y flores: como en las gentes: y siempre de diversa manera: y asi en un lugar como en otro. Lo mismo en los puertos y en las aguas. (Columbus and Las Casas 1989, 172–74)

——————

[He saw coming toward him a great stream of very pretty water that descended a mountain and made a great noise. He went to the river and saw shining in it some stones with gold-colored spots on them, and he remembered that in the Tagus River, in its lower part, near the sea, gold is found, and it seemed certain to him that this one should have gold. And he ordered certain of those stones collected to take to the sovereigns. While he was thus occupied, the ships' boys shouted that they saw pine groves. He looked up toward the mountains and saw them, so large and admirable that he could not praise [sufficiently] their height and straightness, like spindles, thick and thin, where he recognized that ships could be made, and vast quantities of planking and masts for the greatest ships of Spain. He saw oaks and arbutus and a good river and material to make water-powered sawmills. . . . He saw along the beach many other stones the color of iron and others that some said were from silver mines, all of which the river brought. There they took a yard and a mast for the mizzen of the Niña. He reached for the mouth of the river and went into an opening at the foot of the cape, toward the southeast, which was very deep and large, and in which there would be room for a hundred ships without any cables or anchors. And the harbor was such that eyes never saw another like it. . . . And he indicates that he had received from seeing it, and even more so from the pine trees, inestimable joy and pleasure; because as many ships as might be wanted could be made there, bringing out their equipment except for wood and pitch, of which a great plenty would be made there. . . . And that always, in what he had discovered up to this point, he had gone from good to better, as well in lands and groves and plants and fruits and flowers as in people, and always of different sorts, and likewise in one place as in another, and the same in harbors and waters.] (171–75)

First, the sound of a surging waterfall leads Columbus and his men to a shining river's mouth whose shiny pebbles remind Columbus that much gold had been found where the Tagus, on the Iberian Peninsula, meets the ocean (i.e., Lisbon). Then the shouting of the ships' boys directs his sight to a pine grove, whose trees, Columbus observes, are perfect for ships' masts.

The panoramic view of the bay offered by Columbus soon pans upward to the forest, where it pauses at the sight of a waterfall and a river. Again he lowers his gaze to the bay, now transformed into what is known as a "natural" harbor, prodigiously endowed by nature for the capitalist to extract surplus value, qualified as "natural" only because we are by now familiar with its counterpart, the man-made harbor. By the close of that day's entry in the *Diario*, the slippage from *natural* (belonging to nature) and *natural* (native) is made explicit as the admiral itemizes the elements of the bountiful landscape—though seemingly deserted—with the peoples he had encountered previously, on other islands. Under his gaze, the thick tree trunks become shipbuilding planks, and the entire, "marvelous" scene re-creates the waterfall as the force that powers the sawmill that processes the wood from the nearby felled trees, which will become the masts and planks for the ships that will transport the natural riches of the Indies to the Iberian Peninsula: gold, slaves, and, perhaps, silver but also more masts and planks, and iron, for the construction of more ships.[13] In so doing, Columbus condenses various monopolies at once in a vision and a practice that both abbreviates and amplifies the scale of transformation from nature to what Moore (2016) has called *cheap nature* in the capitalist quest for scalability. Here, I see the monopoly of the empresario (as businessman) over nature-power in specific places and the monopoly over the legitimate use of violence by the empresario (the standard-bearer for the sovereign) to constitute and defend the first monopoly over running, powerful water in the Indies.

The admiral's gaze had material consequences on November 25, 1492, with repercussions one month later. As he, and Las Casas, tells it, the problematic *Niña* received a new yard and a mast from that pine grove. Then the caravels went three for three on Christmas Day. As noted above, first the *Pinta* received a new rudder and a new rigging and was recaulked on the Canary Islands in August. Then the *Niña* received a new yard and mast in late November. By December 25, 1492, the relationship nature-native-capitalist comes full circle when the *Santa María* runs aground on some shoals. What follows is one of the most iconic scenes of the *Diario*, replete with natives crying at the predicament of Columbus and his men: the admiral is received by the king of the village, and the newly minted "don" and the native king exchange a "picturesque moment" over the gift of gloves and a shirt (Columbus and Las Casas 1989, 284).[14]

The day after Christmas, Columbus decides to dismantle the *Santa María*, down to the last lace-end, plank, and nail (*agujeta, tabla y clavo*), in order to build a fort and a moat. With these building materials stripped from the caravel, Columbus also leaves behind bread (*pan*), wine (*vino*), seeds (*simientes*), a rowboat (*barca de la nao*), a caulker (*calafate*), a carpenter (*carpintero*), a gunner (*lombardero*), and a barrel maker (*tonelero*) (Columbus and Las Casas 1989, 289). Fully dismantled and accounted for, the *Santa María* would have left behind a mast and yard, similar to those taken for the mizzen for her sister ship, the *Niña*, from the pine grove found on that other Caribbean island on November 25. The *Santa María* must have also left behind a rudder, just as the *Pinta* did on Gomera, only days after they had set sail from Palos in August 1492.

Thus wood metonymically fulfills the cycle for conquest when understood as a state of emergency. It is harvested for the ship; another ship is repurposed for a fortress; and the wooden fortress, though one that Columbus did not feel was materially necessary, completes another stockade on Hispaniola.

> No porque crea que aya esto menester por esta gente: porque tengo por dicho que con esta gente que yo traygo sojugareia toda esta isla: la qual creo que es mayor que Portugal y mas gente al doblo o mas son desnudos y sin armas y muy cobardes fuera de remedio. (Columbus and Las Casas 1989, 288)

[Not that I believe it to be necessary because of these Indians, for it is obvious that with these men that I bring I could subdue all of this island, which I believe is larger than Portugal and double or more in number of people, since they are naked and without arms and cowardly beyond remedy.] (289)

In the *Diario*, Columbus argued that the fortress was *symbolically* necessary so that the Indians would learn to obey the Catholic monarchs, "with love and fear" (*con amor y temor*). Thus the material and symbolic conditions for the state of emergency become recycled and repurposed from island to island.

Perhaps it is not that far-fetched to think that those three caravels—the *Niña*, the *Pinta*, and the *Santa María*—were constructed in part from fine cedar logged in Madeira, which supplied wood to Spanish as well as Portuguese shipwrights, the same island chain where Columbus had learned his trade from his in-laws years before. The wood of Madeira is but one example of scalability. When Columbus describes the great girth and height of the trees he found in the (West) Indies, he is making an oblique comparison to the lumber from the Madeiras, prized for its strength and the height and width of its planks. It was the naturally larger scale of the planking from Madeira that had allowed Portuguese and Spanish shipwrights to build both larger ships and larger fleets that could withstand ocean crossings (Perlin 1989, 45–46). This "scalability" of seafaring empires was and is inextricably linked to those measurements and quantities afforded by nature; there is no escaping the metonymic relationship.

A metonymy similar to that of Madeira's wood was later explored by Bartolomé de Las Casas in *De Thesauris* (1564), which means "On Treasures." However, instead of tracing the trajectory of wood from one island to the next across the Atlantic, in this treatise Las Casas mapped the looting of precious metals and gems stolen from graves from one conquest to the next on the American continent. In so doing, Las Casas narrated the story of conquest in *De Thesauris* as an accumulation of moral and material indebtedness accrued by the conquistadors and, by extension, the crown that had received the quinta real from each venture it had chartered. The legacy of *De Thesauris*—its haunting quality in the work of José de Acosta—is argued in the next chapter.

Telling the story right, from the beginning, may be one of the main concerns of the scholars and activists who advocate for the use of the term "Capitalocene" rather than Anthropocene, as the latter retains the binary humanity/nature, which is, in and of itself, a product of the relation of capital, power, and nature, understood by Moore (2016) as an organic whole that emerged and expanded rapidly as a new world ecology during the long sixteenth century.[15] For Justin McBrien (2016, 119), who has advocated in favor of his neologism Necrocene, with reference to Mbembe's concept "necropolitics," the use and abuse of the term "Anthropocene" "reinforces what capital wants to believe of itself: that human 'nature,' not capital, has precipitated today's planetary instability." Paradoxically, the personification of capital in Moore's *Capitalism in the Web of Life* (2015) itself underscores the dehumanization of the human and the reification of nature, that is, how the social relations of capitalism arise out of and reproduce a distorting distinction between humanity and nature.

For proponents of the use of "Capitalocene," then, the employment of "Anthropocene" places an emphasis on the burning of fossil fuels and tells the history of capitalism's effects on externalized nature beginning in eighteenth-century England, with the Industrial Revolution, yet the origins of the world-ecological system under capitalism are to be found instead in the long sixteenth century. Rather than locate the origins of capitalism in the industrial capitalism of England in the late eighteenth century, proponents of the Capitalocene point to the partnerships between Iberian monarchs and Italian capitalists, mapped out in the first chapter, and their dependence on Indigenous and African slavery beginning in the fifteenth century. Thus the transatlantic space and its islands were the incubator of our current world ecology, as the logic of capitalism transformed its places and its peoples into metonyms for cheap nature. The analysis of material metonymy allows us to see the (et/est) links between beast, sovereign, and, I would add, capital.[16]

As Las Casas wrote his history in counterfactuals, that is, the subjunctive mode, Madeira was suffering a profound transformation, not unlike the destruction wrought by Perestrello's rabbits. By 1660, Madeira was the "island of wood" in name only; between the deforestation, to build ships and fuel sugar mills, and the depletion of the topsoil

for cane cultivation, Madeira had become barren, much as Porto Santo
had been devastated, centuries earlier, by one very pregnant rabbit and
the conquistadors in tow. In the *Historia de las Indias*, Las Casas—
though on the surface espousing the providential design of the "discovery" from the opening lines—implicates the contingency of the
New World discovery: were it not for Christopher Columbus learning his trade from his in-laws, Portuguese-Lombards, on Madeira and
from similarly hyphenated compatriots in Mina, indeed, were it not
for the Portuguese ventures, family connections, and the like, Columbus would not have "discovered" America. In yet another exploration
of the counterfactual, Las Casas contended that had Columbus been
blessed by the foresight to witness the destructive effects of his discovery, he never would have set out for the Indies.[17] From his editorial and
marginal notes in the *Diario*, or his narration of Columbus's atrocities
in the *Historia de las Indias* and the *Brevísima*, we may conclude that
Las Casas was being disingenuous or ironic, at best, with regard to Columbus. The counterfactual, however, posed to contemporary readers
who are blessed with such hindsight, frames the question in a different
light: What would you do, and *do* you do (now), with that knowledge?

IN CHAPTER 2 I explored how scripts, laws, and contracts produced
by Spain's bureaucracy aimed to guide its agents as they navigated contingencies with the imperial telos on the horizon. Navigating these con-
tigencies was no easy task, as the happenstance of each enterprise in the
short term had to be construed within the long-term providential design. The imperial itinerary was made all the more complicated by the
desire, expressed in the law, for Indigenous consent, on the one hand,
and the recognition, on the other, as outlined in contracts, of the agents'
profit motive, which often led to a preference for coercion, and thus
enslavement. At the same time, the contracts between conquistadors
and the crown preemptively accounted for Indigenous slaves while the
Requerimiento ostensibly offered Indigenous interlocutors a choice between enslavement or subjecthood.

In this chapter, I have traced the wake left by letrados when faced
with the difficult but not insurmountable task of navigating the contradictions comprised by the official documents of the imperial rep-
ertoire. Following his experience with implementing the legal fictions

in the Darién, including branding the skin of Indigenous persons, Fernández de Oviedo wrote the *Claribalte* as he waited for a ship to return to Spain. The reproduction of the conventional tropes of chivalric fiction in the *Claribalte* replicates the repeated performance of the *Requerimiento*—first brought to the New World by Pedrarias Dávila and his cohort—and its claims to universality while erasing the specifics of the New World context. It is a fitting inaugural work for an author who will continue to replicate the documentation and erasure compulsion as royal historiographer, even as he silenced his less savory roles in the Darién.

Another letrado of the empire, Las Casas, visibly struggled with his role within the colonizing apparatus. In his *Historia de las Indias*, the compulsion to document all aspects of the conquest over decades while also editing the *Diario a bordo* by Christopher Columbus allowed him to reflect on individual and corporate personhood and levels of accountability, including his own. Rather than attempt to reconcile the accounts written with different temporal horizons in an effort to preserve the individual merchant or conquistador's credit, Las Casas bridged the gap between daily log and historical arc by dabbling with the counterfactual in his treatment of Columbus. The writing of history in the subjunctive mood will increasingly dominate his approach to the relations between the Indies and Christian conscience, one against which José de Acosta would rail long after the Dominican friar's death.

CHAPTER FOUR

The Specter of Las Casas in the Political Theology of José de Acosta

They said that the Pope must have been drunk when he made
[the Donation], because he gave away what was not his to give,
and the King, who had requested and accepted the gift, must be
some kind of mad man, because he asked for things that belonged
to others.

—Martín Fernández de Enciso, *Suma de geografía*

The exchanges produced by venture capital in the Spanish Conquest
depended on an understanding of love that privileged the metaleptic
use of *amor* and *interés* as synonyms, seemingly without contradiction.
Reception of this ideology defined the political and religious thought
of Bartolomé de Las Casas and José de Acosta, whose influence was
felt beyond the Spanish-speaking world among their contemporaries
and subsequent generations.[1]

Both authors have been acclaimed as precursors to liberation theology and "teología india" (Indian theology), movements that respectively privilege the poor and Indigenous cultures' self-understanding.[2]
This chapter examines the related concepts of freedom and salvation
as professed by the Dominican friar Bartolomé de Las Casas and their
reception and criticism by the Jesuit José de Acosta. Although the authors shared the conviction that the Spanish invasion of the Americas
had been illegal, their views on free movement for the purposes of
trade and evangelization methods, the two loopholes to jus gentium

explored in chapter 1, could not be further apart. As these moral and legal exceptions are related directly to the *state of exception* in the Indies, which I examined in the first two chapters, these authors' respective thoughts on these two exceptions define their political theologies. For both authors, it was not enough to define one's position within the circular logic of the law. The prescriptions for converting the Indigenous peoples to Catholicism made by Las Casas and Acosta were inextricably tied to their understanding of the Indigenous subjects' self-knowledge and knowledge of the world.

Were conquista and its discontents "haunting" the Indies held by the Spanish Empire? Acosta belonged to the writers of the post-1573 generation for whom conquista, following the prohibitions of the use of the term, ought to have been a memory.[3] It would be impossible to imagine Acosta writing *De procuranda* (1589) and *Historia natural y moral de las Indias* (1590) without the written work and activism of Las Casas in mind. Yet not once does Acosta mention Las Casas by name in *De procuranda*, even though he polemicizes throughout his work with the entire body of Lascasian thought, especially from the latter years. Though Las Casas belonged to an earlier generation of missionaries in the Americas and had died four years before the young Acosta would receive his order to travel to Peru in 1570, the Dominican friar cast a long shadow over the Jesuit's mission and livelihood. When Acosta arrived in Peru he found that the new viceroy Francisco de Toledo (1515–82) had recently upended the crown's strategy regarding the Vilcabamba insurgency by choosing military invasion over diplomacy. The last Inca of Vilcabamba, Tupac Amaru I, had been executed in the public square in Cuzco only months before Acosta set foot in Callao in April 1572. On the ideological front, Toledo scoured the Inca centers of power for native informants whose histories could attest to Inca tyranny. His efforts attempted to counteract the proponents of self-rule based on native elite structures, such as the Incas or the curacas, without the secular apparatus of the colonial state.

The advocacy of Dominican friars, such as Domingo de Santo Tomás, on behalf of the curacas of the Mantaro Valley in the 1560s had received theoretical support from Las Casas in *De Thesauris* (1563), which sought remedies and remuneration on behalf of the Indigenous peoples for the lives and treasure they had lost during the Spanish invasions.[4] Las Casas's *Doce dudas* (1564) asserted that the damage done to the Indies was irredeemable and that Spain's monarchs and its people would be forever damned unless they retreated from the Americas. Supporters of Las Casas were dispersed among faculties of theology and Indigenous languages in Lima and among Dominican, Jesuit, and Augustinian friars in the highlands. The advocacy for self-determination by this Lascasian network was opposed by *extirpadores de idolatrías* (extirpators of idolatry) such as Martín de Murúa (1525–1628) and Cristóbal de Albornoz (1530–?).

José de Acosta established close ties with the viceroy and the ecclesiastical hierarchy in Alto and Bajo Peru. Soon after his arrival in Lima, Acosta traveled throughout Cuzco, Juli, Arequipa, La Paz, Chuquisaca (Sucre), and Potosí on an itinerary and timetable that largely coincided with Viceroy Toledo's informaciones and ordenanzas. On his return to the Ciudad de los Reyes (Lima), Acosta participated as a *calificador* (judge on theological matters) in the Inquisition trial against three Dominicans, Francisco de la Cruz (d. 1578), Pedro del Toro, and Alonso Gasco, and a woman, María Pizarro.[5] De la Cruz was burned at the stake for heresy; his citations for apostasy against the faith included his support for polygyny among laymen, the abolition of celibacy for priests, and advocacy for women's priesthood and the self-determination of the Indigenous peoples of Peru. De la Cruz had occupied a high post in the ecclesiastical hierarchy of colonial Peru; his beliefs, his apostasy, and his downfall must have had a profound effect on Acosta, who would later warn fellow and future missionaries in *De procuranda* of the corrupting influence of native practices on the spiritual and physical health of priests.

Acosta saw the influence of the devil in native practices, which, uncannily, mirrored Christian sacraments (such as anthropophagy and the Eucharist) and were referred to broadly as examples of *simia dei*, the devil's mimicry of God.[6] His influence on the decisions and catechisms developed during the Third Council of Lima (1582–83) are widely documented.[7] The catechisms emphasized exploration of the native interlocutor's conscience for details of illicit sexual acts (especially sodomy, fellatio, incest, and polygyny) and knowledge of accounting and narrative practices with knotted cords (*khipu*) for extirpation campaigns.[8] By promoting the sacking of tombs and areas of ancestor worship in order to confiscate the narrative khipus, Acosta pushed back against Las Casas's arguments that all objects associated with native burial practices belonged to the Indigenous and their descendants. Soon after his participation in the Third Council of Lima and a brief sojourn in Mexico, Acosta returned to Spain where he continued to scale the hierarchy in his order and in academic circles. At his death in 1600 he was the superior of the Society of Jesus in Valladolid and rector of the Jesuit College in Salamanca.

More than any other humanist and religious thinker who preceded him, including Francisco de Vitoria (1492–1546), Juan Ginés de Sepúlveda (1490–1573), and Domingo de Soto (1494–1560), Father Acosta asserted within the parameters of his faith and reason that spiritual enterprise was a comparable and inseparable handmaiden of free movement for trade. His position on the role of trade in evangelization could not have been more different from the project of Las Casas. Throughout his ecclesiastical career, first as an ordained priest and later as a Dominican friar, protector of the Indians, and bishop of Chiapas, Las Casas had firmly believed in the positive influence of Iberian peasantry and tradesmen in securing the trust of Indigenous interlocutors; he had little faith in capital and its agents.[9] His failed settlement in Cumaná (Venezuela) and his ventures into evangelization in Verapaz were exemplary of peaceful contact for his supporters and detractors alike. Las Casas thus established an important precedent in the discourse of evangelization: first, any discussion of method had to consider the Indigenous habitus prior to contact; second, Indigenous reception of the faith could not be divorced from an honest and detailed accounting of the violent methods used by missionaries and conquistadors and of the injuries sustained by the Indigenous subjects in these encounters. Thus his treatise on the only way to evangelize Indigenous subjects envisioned and inaugurated the practice of merchants and missionaries entering Indigenous lands unarmed and willing to assume martyrdom.

The *Historia de las Indias* and its offshoot, the *Apologética historia sumaria*, complemented *De unico vocationis modo*'s treatise on evangelization by contesting the Aristotelian argument in favor of natural slavery on several rhetorical fronts. As Hanke (1974, 124–25) has observed, Las Casas had a mercurial relationship with the philosopher. If in 1518 Las Casas denounced Aristotle as a heathen philosopher whose beliefs on proper government had no bearing on Christendom, the

Apologética would argue that Indigenous societies displayed the characteristics of civilized societies by referencing the categories employed by Aristotle in his *Politics*. To Juan Ginés de Sepúlveda, the learned humanist and translator of the *Politics*, Las Casas showed little deference, arguing that Sepúlveda had not understood Aristotle's doctrine of natural slavery, which was, in any case, irrelevant to the Indies since only a small number of individuals in *any one* society but not entire *peoples* could be considered natural slaves. Moreover, the Indigenous peoples could not be enslaved under the civil regime envisioned by Aristotle, as the Spanish had no just cause for war against the Indians, who had every right to defend themselves against invaders. At the same time, as Rabasa (1993, 164–79) contends in *Inventing America*, Las Casas later undermined the conventional alignment of binary oppositions (civilized vs. barbaric, Christian vs. heathen, sheep vs. wolves) in the *Brevísima relación* and the *Apologética*, thus questioning the relevance of binary thought and the Aristotelian paradigm for Christianity in the Americas.

Do all these references to Aristotle make Las Casas an Aristotelian thinker, as O'Gorman argued?[10] His ongoing and evolving dialogue with Artistotelian thought throughout his life suggests that Las Casas felt compelled to engage with Aristotle but was not an Aristotelian per se. Confronted with arguments based on ideas of natural and civil slavery at every turn in the justification of the conquest in the Indies, Las Casas whittled away at them for over fifty years, using all the rhetorical tools at his disposal. This included arguing within the Aristotelian system and outside it, depending on the project at hand.

Las Casas learned from his enemies. He would argue, forcefully, that the Indians could learn many things from the invaders, including the rationalization of the irrational. From Fernández de Enciso's *Suma de geografía*, Las Casas discovered the potential of Indians' voices as an original source that resisted the framework of imperial apologetics within which they were presented. Fernández de Enciso is the sixth conquistador depicted in Guamán Poma's allegory of the conquest: he is the conquistador, author, and lawyer on the poop deck, the figurative tailwind of the *empresa*, emblem but also enterprise of the Spanish Conquest.

In the *Suma de geografía*, Fernández de Enciso cited the pronouncements made by Indigenous peoples of the Cenú against the

Requerimiento as examples of their unruliness, their lack of respect for king and pontiff, and their spiteful resistance to Spanish dominion.

> Respondiéronme que en lo que decía que no había sino un Dios y que éste gobernaba en el cielo y la tierra y que era señor de todo, que les parecía bien y que así debía de ser, pero que en lo que decía que el papa era señor de todo el Universo, en lugar de Dios, y que él había hecho merced de aquella tierra al rey de Castilla, dijeron que el papa debía estar borracho cuando lo hizo, pues daba lo que no era suyo, y que el Rey, que pedía y tomaba la merced, debía ser algún loco, pues pedía lo que era de otros, y que fuese allá a tomarla, que ellos le pondrían la cabeza en un palo, como tenían otras, de otros enemigos suyos . . . y dijeron que eran señores de su tierra y que no había menester otro señor. (Quoted in Las Casas 1994, 2020)

> ───────────

> [They answered me that, as for my saying that there was but one God, and that this God ruled heaven and earth and was Lord of all, this seemed all well and good, and this is how things must be. But when it came to the pope being lord of the universe, in the place of God, and that he had granted the favor of bestowing that land on the king of Castile, they said that the pope must have been drunk when he did that, since he was "bestowing" something that was not his, and that the king who requested and accepted the favor must have been somewhat mad, since he was requesting something belonging to someone else, and that he should come here and [try to] take it himself so they could drive his head on a stake as they had with other enemies. . . . And they said that they were lords of their land and that there was no need of another lord.]

In the same passage, Fernández de Enciso also accused the Indians of committing suicide out of spite, to deny Spanish access to labor, and rejecting forced baptism. As seen in chapter 3's anecdote about the pregnant rabbit of Madeira lifted from Barros's *Ásia*, Las Casas cites from Fernández de Enciso's *Suma de geografía* almost verbatim. Limiting his commentary to a reiteration of the arguments and questions raised by the men of Cenú, Las Casas concludes, sardonically, that this bachiller, believer in legal fictions, must have created what could only amount to a fictitious response.[11] For in the space of an hour and with no knowledge of each other's languages, how could they have discussed pontiffs, donations, monarchs, and the Trinity? And yet, as with the eloquence of Hatüey, the defiant Indigenous chief, or the mothers justifying infanticide in both the *Brevísima relación* and the *Historia de las Indias*, Las Casas does not deny the substance of the claims allegedly voiced by the people of Cenú. In other words, Las Casas admits the possibility that spite was what drove the Indigenous responses to the *Requerimiento*; indeed, his economy of retribution factors in the original action of the conquistadors and evangelizers (i.e., Indigenous *despecho* [spite] for the conquistador's amor).

Las Casas does not deny these motivations; rather he argues that rational, caring human beings would not have reacted in any other way. It has been widely noted that suicide in the public square is the greatest example, and the most feared act of defiance, of the empire's limits on regulating biopower.[12] Suicide introduced a radical alterity into that equation and, despite Fernández de Enciso's protestations to the contrary, articulated a rational rejection of Christianity as it was practiced in the Americas by its zealous proponents, especially the labor and spiritual exchanges of the encomienda system that placed a premium on both the labor and the souls of potential neophytes.

With *De unico vocationis modo omnium gentium ad veram religionem* (ca. 1538), Las Casas aimed to widen the breach between the general partners of the Conquista—church and state—and to reintroduce missionary work as a labor fraught with moral and physical peril. His ideal missionary would not count on armed men to guard his life; thus armed men would have no place in the social order of Tezulutlán, Guatemala, which Las Casas renamed Tierra de la Verapaz (Land of True Peace), where he intended to reproduce apostolic Christianity in the Indies. Thus he proposed entering Tezulutlán, a place known as *la tierra de guerra* (the land of war) by the Spanish colonizers, without any means of physical defense even though their efforts were being met with armed resistance from the Mayan peoples there.

As shown in chapter 2, Indigenous peoples of designated tierras de guerra were living a social death, as they were already accounted for as slaves-to-be-captured. For the Dominican missionaries in Verapaz, there were no guarantees of success and no temporal horizons to enforce timely conversion. At the same time, the haste of the conquistadors and their leaders to gain dominion over native lands and labor had created malicious delay in the true conversion of the Indigenous. For Las Casas, this damage to others was irreparable. Robbed of their time to repent, the Indigenous saw their opportunities for salvation cut off,

> sed potissime tempus vitae, quo necessario indigent ad fidem suscipiendam, baptismum et paenitentiam, tollantur. . . . Ergo maximum peccatum inter peccata contra proximum committit, qui causam dat illius perditionis. (Las Casas 1990, 508)

> [especially for having robbed them of their lifetimes, which would have been enough to receive the faith, baptism, and repentance. . . . Thus they commit the gravest of all sins against another, for they have been the cause of the other's perdition.]

Rooted in Aquinas's understanding of time as a thing of God, the theft of the time and means of conversion—unlike material and spiritual goods—had no remedy or satisfaction. Perhaps the only satisfaction would be the knowledge that the thieves had committed a mortal sin that could not be absolved, since no act of reparation could adequately compensate the aggrieved for the time lost that might have been spent in a state of grace: "sunt ergo rei supradicti omnes totius damnationis omnium eorum" (so that all the aforementioned men are accused of damnation) (Las Casas 1990, 508). Only sins against the Holy Spirit are unforgivable, but by raising the specter of the impossibility of accounting for the lost time in the state of grace of another, Las Casas approaches an aporia in the Catholic economy of faith and repentance. From Las Casas's economy of time and repentance, we can infer that absolution for encomenderos, and ultimately the sovereign, was impossible. It was an impossible possibility that went against the doctrine of the church, as all sins except for those against the Holy Spirit are forgivable, but it nonetheless existed by fiat, de facto, much like the Spanish Empire was said to exist in the Indies, because a handful of men, the preachers who refused to give the encomenderos absolution, had willed it so.

Moreover, Las Casas notified, that is, *required*, the encomenderos of Verapaz and beyond to remain entirely removed from dialogue with Indigenous peoples, to avoid a legal loophole for the use of justified violence against those who remained or would remain nonbelievers. In addition to the free movement for trade and evangelization exception to jus gentium, which could be "defended" by violent means, another commonly held "suitable and legitimate title" of appropriation was the right of Christian states to send armed men to defend Christian neophytes within communities of nonbelievers. As the full title of his treatise suggests, *De unico vocationis modo* omnium gentium *ad veram religionem* (emphasis mine), Las Casas struggled with the "protected" status of Indigenous Christian neophytes within larger communities of Indigenous nonbelievers. His proposed method for conversion thus foreclosed the legal loophole to jus gentium that would have permitted a "divide and conquer" approach. Instead, where Las Casas turned to trade, he did so to place it at the service of missionary work within consensus-building forms of self-government in Indigenous communities. Accordingly, when he attempted to put his ideas into practice in Verapaz, Las Casas enlisted Indigenous Christian merchants who would continue trading with the *indios de guerra* (warring Indians) while singing of their faith in Maya. It was only when their professions of faith were welcomed, and the community reached a consensus to invite missionaries to preach, that Spanish missionaries would follow Indigenous traders into this tierra de guerra. The emphasis on "omnium gentium" underscored both the universal ambitions of his project and its inherent paradoxes: how to maintain the cohesion of the gentium while undergoing the fracturing and denial of self that is inextricably part of the conversion process.

Both the *Brevísima relación* and the *Historia de las Indias* offer an overwhelming casuistry that defends the real political subject in each and every case of the attempted subjugation of Indigenous communities through individual conversions that sidestepped consensus-building collectives in pursuit of the exception. As Giorgio Agamben has shown in his analysis of Paul's letters, messianic time is defined by the suspension of the law, of the exception *as* law, when paradox reigns and our understanding of "the people" becomes fragmentary. For Agamben (2005, 57), the political legacy of Pauline messianism (the *when* and *how* of conversion to Christianity) is the remnant, "that

which can never coincide with itself, as all or as part, that which infinitely remains or resists in each division, and, with all due respect to
those who govern us, never allows us to be reduced to a majority or
a minority. The remnant is the figure, or the substantiality assumed
by a people in a decisive moment, and as such is the only real political
subject." This real political subject, whom the *Requerimiento* created
while denying its very right to exist, also offers another frontier, in
consciousness, to which imperial power would stake a claim.

Early in his writing career, Las Casas attempted to avoid exploit-
ing the exceptions and took pains to recognize the existence of the real
political subject in increasingly fragmented Indigenous communities as
he participated in Christianity's futurity by proposing his own model
for evangelization. However, he increasingly found that he could not
address the future without making amends for the past. If those who
argued in favor of Spain's legitimate titles of war against the Indians,
such as Juan Ginés de Sepúlveda, were indeed a threat to the welfare
of the Indians, the jurists, with their exceptions to jus gentium, such as
Francisco de Vitoria, could prove equally pernicious. These exceptions
formed part of the strategy used to expand empire daily: the necessity
for an armed "defense" of traders, missionaries, and Christian neophytes through the production of (supposedly) universally held values.
As the Laws of the Indies increasingly claimed to outlaw Indigenous
slavery, on the one hand, while upholding the protected status of the
three aforementioned groups and their armed "defense," on the other,
Indigenous bondage became increasingly illegible. Its invisibility has
continued to haunt historiographic accounts of the Spanish Empire,
as Reséndez in *The Other Slavery* (2016), Nemser in *Infrastructures
of Race* (2017), and Van Deusen in *Global Indios* (2015) have argued.
According to Las Casas, thus, by the 1560s, the discursive threat to
Indigenous communities lay less in those arguments wielded in favor
of a "just war" against the Indians than in those habitual practices of
profitable violence that had already claimed the lives and wealth of the
Indigenous peoples of the Americas through recourse to the exception.

EXCEPTION TO THE EXCEPTION

Acosta's project to free Indigenous subjects was surely no less earnest
than that of Las Casas. It is thus hardly surprising that Acosta would

model his *Historia natural y moral de las Indias* and *De procuranda* on the influential works of the Dominican friar. However, the similarities between their works were limited to format: a Latin treatise on evangelization accompanied by a history of the Indies that examined the Indigenous habitus before and after the conquest. Though their projects had a similar approach, their ethos could not be further apart in the culmination of comparative ethnology within Aristotelian categories in Spanish letters.[13]

Acosta and Las Casas held similar positions on the Papal Donation. As argued by the School of Salamanca, especially the articulation of legal titles of appropriation by Vitoria, the Papal Donation to Spain and Portugal did not constitute a legal transfer of Indigenous dominion over their communities and land to the Iberian monarchies (Schmitt 2003, 113–20). Vitoria contested rights of appropriation based on imperial world domination, papal world domination, the right of discovery, the rejection of Christianity, the crimes of barbarians, the free consent of Indians, and divine (providential) donation. He called these seven titles "unsuitable and illegitimate" (*tituli non idonei sec legitimi*). However, Vitoria's seven "suitable and legitimate titles" (*tituli idonei ac legitimi*) provide an opening for inquiry into the assumptions of hegemonic, universalist thought: the right to free movement for trade (*jus comercii*), the right to preach the faith (*jus propagandae fidei*), the right to protection for Christian Indians by papal mandate (*jus mandati*) as an intervention against tyranny (*jus interventionis*), the right to free choice (*jus liberae electionis*), and the right to protect one's allies or associates (*jus protectionis sociorum*).[14] Thus, though Vitoria contested Sepúlveda's arguments for Spanish titles to the West Indies that were based on Spain's purported cultural superiority (i.e., the Aristotelian argument in favor of subjugating weaker classes of humanity) or on the imputed crimes of Indians against others, such as cannibalism or human sacrifice, when confronted with particular cases from the Indies, Vitoria's arguments often justified the Spanish presence in the Indies through one of the legal titles, as Bentancor (2014) has contended.[15] Vitoria also agreed with Sepúlveda on the matter of free movement for the purposes of trade and the dissemination of the faith. Opposition to the free passage of traders and missionaries was a just cause of war (*causus belli*). Moreover, Vitoria defended Spain's right to intervene on behalf of those Indians who had converted to Christianity. Thus the presence of Christian neophytes in Indigenous communities effectively circumscribed them within the tautology of legitimate violence. Yet, as seen in the previous section, questions on the *time* and *means* of conversion had brought into focus the loopholes through which universal imperatives, such as the dissemination of Christ's news in the known world, could override local beliefs and forms of government (jus gentium). In this way, messianic time and place staked their own claim in the consciousness of each potential neophyte.

What then of jus gentium when the people were caught up in the process of their own destruction? Vitoria's position on free movement for trade and missionary work never engaged the question of the time and means of conversion.[16] While Vitoria did not consider the *moment* or *means* of conversion as part of his broader arguments on the legitimate use of force by foreign powers, these were the considerations at the heart of most approaches to missionary work in the Indies, including those of Las Casas, Soto, and Acosta. Also, although Vitoria had called for restitution for injuries suffered by the Indians in his *Letter to fray Miguel de Alarcos*, citing the Peruvian case in particular, he did not go as far as Las Casas, whose calls for remedies for the time lost by Indigenous subjects addressed the theological economy of the conquest as one in which the circulation of God's grace depended on sinful consciences and actions taken through three scandals: the conflation of cupiditas and caritas, the theft of time for salvation, and the spoliation of native graves. As explored in chapter 3, Las Casas did not believe that this economy was limited to the boundaries of the Spanish Empire and leveled comparable accusations against the Portuguese Empire in Africa and the East Indies as well.

For Las Casas, free movement for trade was little more than an excuse for armed men to elbow their way into foreign ports.[17] He tells the story of King Manuel of Portugal's armada that had been sent to India in 1500, with reference, once again, to João de Barro's *Ásia* (Las Casas 1994, 1227). In the first section, Las Casas's narrative voice conforms to the metaleptic habitus of venture capital, including approval for the division of "spiritual" and "material and temporal" agents among the Franciscans and mariners in the Portuguese expedition. Everything, notes Las Casas, was done according to canon law, including the necessary "requerimientos" with the formula that would become even more notorious in the West Indies some years later. The Portuguese requisition made reference not only to Christ's "caridad y ley de amor" (charity and law of love) but also to the need to respect "comercio o conmutación, que es el medio por el cual se adquiere y trata y conserva la paz y amor entre todos los hombres, por ser este comercio el fundamento de toda humana policía" (commerce or exchange, which is the means by which humanity procures, fulfills, and conserves peace and love, for commerce is the foundation of human civilization) (Las Casas 1994, 1227).[18] In this way, Las Casas recognized the ethical appeal to the argument in favor of free movement for trade and seemed to validate it, as the exchanges of material goods would restrain humanity within the bounds of civilized discourse.[19] Nonetheless, Las Casas introduces the caveat, "pero con que los contratantes no difieran en ley y en creencia de la verdad que cada uno es obligado a tener y creer de Dios, que en tal caso les pudiesen hacer guerra cruel a fuego y a sangre" (as long as the parties do not differ in religion and belief in the truth, for every one is obligated to have and believe in God, in which case they could wage the cruelest war) (1228). Las Casas opposes each individual conscience against the threat of all-out war (*guerra cruel a fuego y a sangre*). The phrasing "no difieran en ley y en creencia de la verdad" establishes a counterfactual; after all, it is *because* these peoples are not believers that they and their lands are pursued with impunity by these commercial enterprises. Once again, Las Casas writes history in the subjunctive mood to expose the absurdity of legalistic formulas.

In the paragraph that follows, Las Casas unleashes the full wrath of his ironic wit, observing that the Indians of the subcontinent received the faith "a porradas" (by blows). Moreover, trade as "the foundation of human civilization" was a misnomer for material exchanges made under duress, for "aunque no quisiesen, habían de usar el comercio y trocar sus cosas por las ajenas, si no tenían necesidad dellas" (even if they did not wish to, they were to trade and exchange their things for another's, even if they had no need for them) (1228). Las Casas concludes by noting that, much like their counterparts on the other side of the Tordesillas line, the Portuguese sought out violent resistance in order to justify the slavery of the Indigenous inhabitants, the other indios. Here we might recall Ferdinand's admonitions to Dávila and his crew to avoid the temptation to subjugate all Indigenous inhabitants of the Darién as indios de guerra, the ramifications of which I mapped in chapters 2 and 3.

As noted in chapter 1, Acosta admired the Portuguese model of conquest. He found significant differences between the Portuguese and Castilian monarchs, and in the reach of their power, as a function of stakeholding in each imperial venture. Acosta shared with Las Casas an acute understanding of the material motivations behind the Iberian conquistadors' actions in the East and West Indies. Unlike Las Casas, however, Acosta admired the cupiditas of the conquistadors and wished that missionaries would be similarly motivated by the promise of spiritual profits (i.e., Christian neophytes) in the Indies. Specifically, Acosta contemplated missionary work among *barbarians*, whereas Las Casas wrote of working among *peoples*. Las Casas envisioned evangelization as an ongoing appeal to the understanding (*intellectus*), following Augustine's famous image of the convert smashing his idols that culminated in the convert's selective rejection of that aspect of his former self. Citing Augustine at length, Las Casas explores the pain of loss when it is done voluntarily, asking his readers:

> Nam sit, ut ait Augustinus in quodam sermone, leve cuique non est dimittere propria et sectari aliena incerta, dimittere quod scias, querere quod ignoras. Quis enim propria sine dolore deseruit aut sine lacrimis dereliquit? (Las Casas 1995, 446)

> [Let us consider also what Augustine says in a sermon: leaving one's goods to follow uncertain and alien things; leaving the known for the unknown is no small burden for anyone. Who abandons his things without shedding tears?]

If abjuration of one's former self and *belongings* is difficult, when done willingly, Las Casas reiterates, the pain of forced loss is incommensurable, a permanent wound to the soul. Though Las Casas was obviously committed to converting Indians to the Catholic faith, his emphasis on the coherence of the Indigenous habitus—whose subjects would not have been denied God's universal salvific grace, as he insisted in the *Apologética historia sumaria*—contemplated losses as part of conversion. Material remedies and spiritual gratification always had to be accounted for, as it was impossible to speak of true conversion without loss, however freely given.

The questions for Las Casas, then, were whether knowledge of the true faith could make up for all the losses suffered by the Indigenous peoples given the reality of Christianity's introduction in the Americas, which was a scandal (*escándalo*). "Scandal" is and was a doctrinal term used by the church to refer to Christian subjects' attitudes or behaviors that lead others to sin. As such, it was understood as a social relation within an economy of grace. One's power over another or the ability of institutions or laws to induce sinful thought or actions had to be accounted for when evaluating the gravity of a committed scandal. In his writing on evangelization in the Indies and the Indigenous habitus prior to the conquest, Las Casas tied the economies of grace and scandal through the social relations between Spanish and Indigenous subjects.

According to Las Casas, reparations were owed to the infidel or the convert for any violent intrusions on the subjective process of the understanding and the will, especially due to scandal. This entailed adjudicating the liability of scandal to a person, a law, or an institution for every instance of scandal, which may be pervasive in all areas of life. Book 6 of *De unico vocationis modo* developed the various material and spiritual remedies required to compensate Indigenous subjects for the injuries they had received, not only due to unlawful invasion and possession of their belonging (naturaleza), but because of the *scandalous* introduction to Christianity they had received at the hands of the conquistadors.

However, to the extent that conversion was itself a disruptive process, Indigenous subjects always stood to lose part of themselves, as a community. Thus Las Casas not only exposed the identitarian problem of Christianity in Indigenous communities as a means for colonizers to exploit the exception to jus gentium but also described its functioning as a wedge wielded to fragment the self's understanding of being a member of a collective, one that could be, indeed, *had to be*, self-inflected for a true conversion to take place. In this way, Las Casas formulated his own metaleptic counternarrative to the conquest, for if he was willing to concede that Indigenous subjects might have benefited from universal salvific grace without a sincere and explicit profession of the Christian faith, the providential design behind European encounters with the Indies was thrown into doubt via this exploration of the counterfactual, thus placing scandalous Christians in insurmountable debt to God and to Indigenous subjects and communities. At the beginning of his career, Las Casas may have viewed in a positive light the feasibility of providing reparations, but by the time he wrote the *Doce dudas*, he could not imagine Christianity in the Americas *with* the presence of European Christians without accounting for pervasive scandal that had incurred net losses, material and spiritual, both to Indigenous subjects and to the people of Spain and their sovereign.

Accounting for remedies and restitutions had been integral to Las Casas's thought since at least 1537 when he preached *De unico vocationis modo* and started putting it into practice in Tezulutlán, the tierra de guerra he preferred to call Verapaz. Throughout that work, Las Casas repeated Christ's injunctions against taking possessions while doing missionary work or receiving gifts from converts because such material exchanges could easily confuse potential neophytes on the true objectives of the mission.[20] Moreover, the prohibition against material accumulation before, during, or after missionary work would also condition preachers to avoid the (con)fusion of caritas and cupiditas. The Dominican's insistence on decoupling trade and evangelization later received the most vociferous response from José de Acosta.

Acosta's latter-day accounting for new souls in the Indies did not reckon with Indigenous losses. Instead, he referred collectively to free movement for evangelization and trade as an activity that merited armed protection, following Vitoria's lead. As Acosta found Indigenous societies *lacking* in general, Indigenous interactions with Christianity and its believers, in whatever guise, would always provide these Indigenous individuals with a net gain in material and spiritual terms, even if they had suffered material and moral injuries. Thus the armed entrepreneurs of Iberia were commendable, admirable even, because they pursued profit with zeal to the ends of the earth. For them to make a profit, these entrepreneurs had to supplement the areas of Indigenous life that were lacking with elements of Hispanic civilization. If only missionaries would be similarly inspired to fill in the gaps of Indigenous spirituality!

In what moral universe would entrepreneurs become models for missionaries? For Acosta's comparison to work, the values of spiritual and material gains would have to be interchangeable credits and debits in the parallel columns of the accounting system developed by humanist merchants such as Benedetto Cotrugli in fifteenth-century Italy. Not only would Acosta rehabilitate the synonymous use of *caritas* and *cupiditas*, which had fallen into disrepute by the influence of Las Casas, but cupiditas would become exemplary for the ambitions of caritas, as seen in this passage from *De procuranda*:

> Iam illud multum mouere nos debet, quòd videmus ad gentes profundi sermonis, & ignotae linguae homines penetrare lucrispe, nec deterreri barbarie inmensa, sed universa mercium gratia lustrare. . . . Augeant tam longam, & periculosam peregrinationem avidissimè suscipiunt, ut profecto admirabile sit omnes pene por utriusque Oceani, omnes finus orbis terraru[m] stationibus nauium Hispaniensium teneri, omnes Indorum *Satrapas* cum nostris mercatoribus, & nautis commercium habere. At qui pretiosissimas merces quaerimus animas Dei imagine insignes, qui lucra non incerta, aut breuia; sed aeterna in coelis speramus, linguae difficultatem, locorum asperitate[m] causamur. (Acosta 1589, 171–72)

> [An argument that ought to stir [the] zeal of [missionaries], is to observe the people in this century who are reaching the unreached language groups and unknown tribes for the hope of becoming rich. They are not scared off by the most aggressive barbarians, rather they risk all to take them their business offers and their wares. . . . All the [men] (*omnes*) and [Satraps, provincial lords] (*Satrapas*) of the Indies now trade with our merchants and our navigators. So, there is no reason, then, why we, who are looking for much more precious goods, that is to say the souls that bear the image of God, and expecting no uncertain or short-term profit, but the eternal heavenly kind, should be discouraged by the difficulty of the language and the places.] (Acosta 1996, 164)[21]

Acosta can draw parallels between profitable violence and missionary work because the equivalence between free movement for trade and evangelization had already been taken for granted by most authors before him, including, notably, Vitoria. However, Acosta's praise for merchants and their pursuit of profits as paeans to be emulated by missionaries offers a vociferous response to his greatest intellectual competitor, Father Las Casas. In an ironic and at times disparaging way, Acosta draws comparisons between what were, to the mind of the Dominican friar, incommensurable moral actions and passions that did not deserve equal treatment under God's law.

Acosta was not one to deny what he called the "excesses of zeal" in the pursuit of material or spiritual profits. Rather he advocated in favor of humanitarian reform and efficiency for a universal economic and spiritual system that he viewed as a done deal. As in the *Requerimiento*'s framing of the Papal Donation as an event of (recent) antiquity, the conquest for Acosta was an event firmly ensconced in the (recent) past. Acosta's importance to our story of the union of love, violence, and interest resides in his popularity among his contemporaries throughout Europe and his intellectual and material legacy in the apologetics of empire. Like the work of Las Casas, Acosta's oeuvre, especially his *Historia natural y moral de las Indias*, was influential among his contemporaries and for posterity. However, unlike the figure of the bishop of Chiapas, there are no institutes or organizations that cherish the memory of José de Acosta as an advocate for Indigenous and underserved communities in Latin American societies, even though Acosta's major works have benefited from more continous publication since the sixteenth century. A comparison of the two thinkers along the lines of their reception of Aristotelian thought thus seems somewhat arbitrary, unless it accounts for their respective positions on the time and means of conversion of Indigenous subjects and the use of violence to "defend" free movement for trade and evangelization. Yet before we turn once more to Acosta's theology of Indigenous liberation, we must first consider his position relative to Las Casas's thought on salvation and freedom.

Defining Liberty

Porque son libres, "Because they are free." Bartolomé de Las Casas repeats this refrain time and again as he condemns the encomienda and the laws of conquest throughout the *Historia de las Indias*. But what can he mean by "libres"? Let us recall Cyprian's (in)famous aphorism *extra ecclesiam nulla salus* (outside the church there is no salvation).[22] Does freedom outside of the church equal damnation? For Gentiles born during periods thought to be governed by natural law (Adam to Moses) and written, Mosaic law (Moses to Christ), their implicit understanding of God would be enough; this understanding would be made manifest if their laws and customs followed the precepts of natural law. However, *after* the coming of Christ and the law of grace, the context of the conquest questions this requirement of implicit faith. What about peoples who had lived since Christ's first coming without exposure to the Good News? Could entire peoples be compared to the savage child of Aquinas's conundrum?[23] Moreover, following a thorough catechism, what rational human being would reject the love of Christ?

These weighty questions, which had become quasi-hypothetical in Scholasticism's heyday when Thomas Aquinas first posited his exemplum of the savage child aquiring reason, became an urgent matter in the sixteenth century (*De veritate*, q. 14, a. 11, ad 1). As written, the formula of the *Requerimiento* may have assumed the rational intelligence of the Indigenous subject, as Muldoon (1980) has proposed. However, Muldoon's proposition is overly generous in his assessment of the options given to Indigenous subjects. After all, refusal to submit to the pontiff and monarchs was met with enslavement and death, punishments worthy of naturally born slaves, defined in the Aristotelian tradition as irrational beings. Does it follow, then, that refusal to submit is a sign of the potential new subject's lack of reason? Can one rationally refuse the gratuitous love of Christ?

As practiced, the *Requerimiento* treated Indigenous subjects as the violators of natural law, the "savages and subhumans" who in the evangelization models put forward by humanists such as Sepúlveda, Fernández de Enciso, and Fernández de Oviedo were headed straight for damnation. Accordingly, saving the otherwise damned through violence was the obligation of every good Christian.[24] None of these authors contemplated the possibility of a rational rejection of Christianity since they failed to articulate a rational Indigenous subject in their respective works. Indeed, the most complex articulations of God's universal salvific grace belong to those authors who seriously engage with the Indians as rational interlocutors.

Francisco de Vitoria and Domingo de Soto engaged in nuanced discussions of time, the law(s), and the possibilities of God's universal, salvific grace.[25] For Vitoria, the horrific means of evangelization in the Indies, its scandal, had spoiled Indigenous exposure to Christianity. Thus Indigenous subjects could not be accused of willfully ignoring Christian precepts. Moreover, Indigenous practices and beliefs displayed a natural understanding of God so that the possibility of salvation remained open to them even if they did not express an explicit faith in Christ.

For Las Casas, God's universal salvific grace was denied to none, not even those outside of the church, including those who had not professed freely their faith in Christ and those who had no exposure to Christ's teaching. Placing limits on God's mercy was the height of arrogance; arrogating knowledge of God's divine plan was an affront to God's universal salvific will.[26] He defended the right of all peoples to make "universal truth claims" in good conscience and to persist in free dissent of competing claims, including the claims of Catholicism that he held as his own (Dussel 2007, 6–7). The ramifications of this radical position were many, calling into question the moral and legal validity of the encomienda system but also, and more important, establishing a defense of Indigenous sovereignty that emphasized not only an explicit expression of the consensus of the people to be governed but also of their "universal truth claims" to enter into a true dialogue with their would-be overlords.[27]

Las Casas traces the origin of the encomienda back to the island of Hispaniola, specifically, to a faulty reading of Isabel of Castile's last wishes for the evangelization of the native inhabitants of the Islas and Tierra Firme. The letter is worth quoting extensively as it makes policy for the Indigenous based on the monarch's understanding of *libre*.[28] As early as 1503, the Spanish crown elaborated the encomienda system for Indigenous neophytes in the New World as a social contract *avant la lettre*. In the letter of Isabel of Castile in response to Nicolás de Ovando (1460–1511), cited in part by Las Casas in his *Historia de las Indias* and in full in the *Memorial de veinte remedios* (1542), the Castilian queen offers an economy of liberty (too much, not enough) that favors productive communication and conversation between her Spanish and Indigenous subjects.

Por cuanto el Rey, mi señor é yo, por la Instrucción que mandamos dar á don frey Nicolás de Ovando, comendador de Alcántara, al tiempo que fue por nuestro Gobernador á las islas y tierra firme del mar Océano, hobimos mandado que los indios vecinos y moradores de la isla Española fuesen *libres y no subjetos á servidumbre*, según más largamente en la dicha Instrucción [de 1501] se contiene, y agora soy informada que, á causa de la mucha *libertad* que los dichos tienen, *huyen y se apartan dela conversación y comunicación de los cristianos*, por manera que, aún queriéndoles pagar sus jornales, no quieren trabajar y andan vagabundos, ni ménos les pueden haber para los doctrinar y traer á que se conviertan á nuestra sancta fe católica, y que, á esta causa, los cristianos que están en la dicha isla, y viven y moran en ella, no hallan quien trabaje en sus granjerías y mantenimientos, ni les ayudan á sacar ni coger el oro que hay en la dicha isla, de que á los unos y á los otros vienen perjuicio. Y porque nos deseamos que los dichos se conviertan á nuestra sancta fe católica, y que sean doctrinados en las cosas della, y porque esto se podría mejor facer comunicando los dichos indios con los cristianos que en la dicha isla están, y andando tratando con ellos, y ayudando los unos á los otros, para que la dicha isla se labre, y pueble, y para que estos mis reinos, y los vecinos dellos, sean aprovechados, mandé dar esta mi Carta, en la dicha razón. (Las Casas 1994, 1341–42; emphasis mine)

———

[Previously, my Lord and I ordered that the *indios vecinos* and inhabitants of Hispaniola were *free and not subject to bondage* in the [1501] Instructions that we sent to Sir Nicolás de Ovando, friar, and knight commander of Alcantara, when he was our Governor of the islands and mainland of the Ocean. I am now informed that, because they enjoy *too much* liberty, they escape and avoid conversation and communication with the Christians. Even when offered wages, they do not wish to work and [instead] live like vagabonds, and thus they cannot be found in order to catechize them so that they may convert to our holy Catholic faith. For this reason, the Christians who are on this island, and live and reside there, cannot find anyone to work on their farms and their holdings. And they [the Indians] do not help them pan and mine for gold on the island,

which is detrimental to everyone. And because we wish for the
aforementioned [Indians] to convert to our holy Catholic faith,
and know its tenets, and because this can best be achieved by [en-
forcing] communication between the Indians and the Christians,
who try to reach out to them, on that island, helping one another
so that the island is cultivated, populated, and fruitful, and so that
the gold there is collected, so that these my kingdoms, and the
neighboring ones, will benefit, I outline my orders in the following
Letter.]

For the sovereign, there must be usufruct in the exchanges between
the Indigenous subjects (in this letter, the natives of Hispaniola) and
the "Christians."[29] Literally and figuratively, according to Isabel's instructions, the land must be cultivated and populated. Refusal to work,
especially when there are wages to be had, is for Isabel yet another
sign of the excessive freedom enjoyed by her Indigenous subjects qua
Christian neophytes. For Isabel of Castile, an excess of liberty could be
defined as refusing to listen to Christian doctrine, resisting residence
among Spanish Christians, and rejecting paid work in the fields or the
mines. Thus the liberty to choose one's residence, means of sustenance,
or interlocutors is, according to Isabel, excessive. Following her contention that Indigenous refusals to work, trade (*tratar*), and cohabit
are detrimental to Indigenous neophytes and Spanish Christians alike,
Isabel proposes remedies to be put into place immediately.

> Por la cual mando á vos, el dicho nuestro Gobernador, que, del día
> que esta mi Carta viéredes en adelante, compelais y apremieis a los
> dichos indios, que traten y conversen con los cristianos de la dicha
> isla y trabajen en sus edificios, en coger y sacar oro y otros metales,
> y en facer granjerías y mantenimientos para los cristianos vecinos
> y moradores de la dicha isla, y fagais pagar á cada uno, el dia que
> trabajare, el jornal y mantenimiento, que, según la calidad de la
> tierra, y de la persona, y del oficio, vos pareciere que debieren
> haber, mandando a cada Cacique que tenga cargo de cierto numero
> de los dichos indios, para que los haga ir á trabajar donde fuere
> menester, y para que, las fiestas y días que pareciere, se junten á oir
> y ser doctrinados en las cosas de la fe en los lugares deputados para

que cada Cacique acuda con el número de indios que vos les señaláredes. (Las Casas 1994, 1342)

———————

[For this reason, I order you, our aforementioned Governor, to compel and force these Indians to trade and converse with the Christians on that island and to work in their buildings, to collect gold and other metals, and to labor in farms and their upkeep for the Christian *vecinos* and inhabitants on that island henceforth, upon receipt of this Letter. And you shall ensure that each Indian receives a wage for each day of work, according to the qualities of the land, and the persons, and the stations you think they should have. [To this end,] order each cacique to keep a certain number of Indians under his charge, to ensure that they work when they are needed, and so that they gather the Indians to listen to matters of the faith in designated places on the feasts and holidays they see fit. Caciques should meet there with the number of Indians designated by you.]

This letter is the first instance of the queen making explicit recourse to indirect rule, that is, keeping existing Indigenous power structures in place to the extent that they can be productive for the imperial enterprise and its various conversions. Isabel recognized the authority of the caciques to keep the conversation going, both to compel Indians' conversion to Catholicism and to enforce their labor for the transformation of the island's natural bounty into commodities. Thus Isabel charges native elites with distributing tasks for the Indians and for ensuring the latter's attendance at catechism. In the last half of the letter, Isabel also makes provisions for the basic living conditions and safety of her new Indigenous subjects, their *mantenimiento*, which, in Marx's terms, would translate as the value of labor power, or the "value of the means of subsistence necessary for the maintenance of the laborer," and his or her reproduction (Marx 1992, 1:270–80).

In the seventeenth century, something along the lines of Isabel's economy of liberty will be construed as "the social contract," whose first definition has been attributed to Hobbes, Locke, and Rousseau. These English and French authors speak of the power and freedom of subjects in relation to the sovereign, articulated in the form of government either by the rule of a monarch or by majority rule, but never in terms of labor or the means of production. At the heart of those enlightened models of the social contract lies the exchange, explicit or tacit, whereby subjects give up an excess of liberty for the rule of the sovereign that provides protection of their remaining rights. What goes missing from the "rights" talk of the Enlightenment is the explicit consideration of labor and time in the exchange between individual subjects and the sovereign. In contrast, Isabel's letter to a local magistrate in the West Indies at the turn of the sixteenth century could not be more explicit, both in the terms she employs to express the economy of her new Indigenous subjects' liberty and in her orders that they be incorporated into the empire's productive regime, as laborers and as Christians in potentia alike. In this lesser-known imperial formula of the early sixteenth century, the social contract emerges as a business contract. In Spanish it is a *negocio*, recalling the Latin etymology for "business" (i.e., busy-ness): the negation of *otium*. It obligates the excessively free to make productive use of their time in the service of empire in return for the benefits of a new political-theological, and explicitly economic, order.

The context of the origins of the encomienda, as a political-theological and economic order, clarifies the contours of Isabel's concept of the social contract that references wages, ethnicity, and catechism. Isabel's formulation, driven by the pragmatic concerns involved in subjugating new subjects, emphasizes another aspect of the social contract that is lost in later treatments. This enforcement of the new political-theological order on Indigenous subjects entails loss of life and recognizes that their new livelihood is one of hardship and fear. The importance of Isabel's acknowledgment of Indigenous resistance to Spanish "benevolence" cannot be overstated. The Indians' refusal to concede defeat exposes the ideological lacunae, in the sense used by Althusser (2001), in the forced reconciliation between caritas and cupiditas.[30]

If Las Casas saw in Isabel's letter her admission, however slight, of a collective, native will to live outside of the Spanish political-theological order, he reserves judgment in this section of the *Historia de las Indias*. Instead, Las Casas charges the Spanish sovereigns with ignorance about the true conditions of their new subjects. In subsequent chapters, Las Casas argues that Isabel's last order to Nicolás de Ovando, her representative, was grossly misinterpreted, in bad faith, by her Spanish subjects in the Indies. However, Las Casas does underscore that *in effect*, "en la realidad de la verdad," the obligation to live in repartimientos was a sentence of slavery in perpetuity: "y así los dio, en la realidad de la verdad, perpetuamente por esclavos, pues no tuvieron libre voluntad para hacer de sí nada o algo, sino donde la crueldad y codicia de los españoles quería echarles" (and so she gave them [the Indians], really and truly, into slavery for perpetuity, for they did not have free will to do anything or nothing with themselves but what the cruelty and will of the Spanish wanted for them) (Las Casas 1994, 1342). Las Casas's protestations about Isabel's, then Fernando's, then Juana's ignorance about these true conditions are made in the conditional mode: had they known the truth, they would have immediately sought to remedy the situation.

The struggle to overcome invincible ignorance cuts both ways, for subjects and sovereigns. Yet when it comes to liability and culpability, Las Casas reasserts the responsibility of the general partners in the conquest who participate in an "advisory" role: "principales autem in delictis sunt precipientes seu mandantes et hi qui dant consilium patrandi maleficium" (for the main guilty parties are those who give orders and those who advise evil) (1990, 514); that is, the main guilty parties are those who are in the position to promote the greatest scandal. The limited partnerships that serve to protect the pooled assets of the larger fund and the more powerful partners do not, as Las Casas asserts, work for the Christian economy of salvation. Increasingly, his patience wears thin with Charles V and Philip II. He replaces his ironic references to the monarchs' ignorance with protests about his increasing despair at the sovereigns' delay, suspected as malicious, in their actions or their failure to remedy the loss of life and liberty of Indigenous subjects.

On Graves and Belonging(s)

By turning to the impossibility of redressing the damages inflicted in the present and the immediate past, Las Casas addressed in *De Thesauris* the coherence of the Indigenous peoples of America, and especially of

Peru, as real political subjects within and outside of the capital flows of empire. The *Doce dudas* served as a supplement in moral and political theology to *De Thesauris*. Described by researchers such as Ángel Losada (1992, xviii–xix) as a largely incongruent treatise written in the last years of his life, *De Thesauris* couples Las Casas's arguments favoring reparation for looted tombs and sacred places and beings (*huacas*) in Peru and Mexico with repeated insistence on the illegitimacy of the conquest of the Indies. Like the *Doce dudas*, *De Thesauris* reflects on the judgments that confessors had to make when hearing the confessions of conquistadors and encomenderos. What if a conquistador confessed to pillaging a tomb? Or an encomendero admitted that he owed back wages to his servants? What if they had waged war "a fuego y a sangre"? Or if they had burdened their Indians with too much tribute? Or worked them to death in the mines? What could be asked of them to make reparations? What would be enough to satisfy the claims of the aggrieved, which included the dead or enslaved? Who was ultimately responsible? Las Casas presented these questions and his response in his last will and testament to Philip II.

De Thesauris opens with the specific, material case of the huacas, their belongings and their ownership.

> Nunc autem queritur an illa pertinebunt indifferenter ad quemlibet qui, uel propia autoritate, uel de licentia Regum nostrorum Hispaniarum siue gubernatorum, nomine regio, partes illas gubernantium, quesierit, foderit, repererit et tulerit, animo sibi retinendi, itaque acquirat dominium earum rerum pretiosarum siue thesaurorum et possit, salua conscientia, retinere. (Las Casas 1992b, 12)

_____

> [The question is whether all this [treasure] shall belong, indiscriminately, to whomever that, or by his authority, or with license by our Spanish monarchs, or by the governers who in their name head the governments of these regions, search [*quesierit*], unearth, find, and remove it for the purpose of keeping it, if they shall aquire dominion over these precious treasures and can, in good conscience, keep them.]

In this section, Las Casas's reference to *Regum nostrorum Hispaniarum* specifies his own sense of belonging—his naturaleza—while representing the embodied history of others' belongings, specifically, their dead and the objects associated with their burials. In the document that keeps his own impending interment on the horizon, Las Casas offers a juxtaposition in autochthony between Peru and Spain: recall the places where "our" versus "their" ancestors are buried. This exercise in juxtaposed autochthony—the sense of belonging that comes from the deep earth (*chthon*), both as grave and as source of abundance that must be "unearth[ed]"—goes to the heart of the contradiction in searching out inhabited terrae incognitae, this is to say, the (ab)use of naturaleza, as explored in chapters 2 and 3, that could both exclude humanity and lay claim to native belongings. For Las Casas, nothing exemplified the alienation of naturales from their naturaleza more than this looting of primitive accumulation.

Even Acosta, when conceding to the conquistadors that their violence behind primitive accumulation could be counted as a form of labor, cannot help but cite a biblical passage in the excerpt of *De procuranda* already explored at length in chapter 1. Acosta's references to the burial plot of Joseph in the Book of Joshua drives home his point about the sovereign's power to distribute land possessed through conquest while raising the specter of just compensation for *belonging(s)*—in the double sense of the possession of objects, on the one hand, and membership in a community, on the other—through burial procured by other, nonviolent means.[31] Widely interpreted as a narrative and thematic hinge between the Pentateuch and the Books of Deuteronomy through 1 Kings and 2 Kings in the Old Testament, the Book of Joshua narrates the fulfillment of the promise of the Mosaic covenant with God through the military conquest of the Promised Land by the Twelve Tribes of Israel. As he lays dying, Joshua reminds the Israelites of their debt to God but also to himself: "I gave you a land on which you have not labored, the towns that you had not built, and you live in them; you eat the fruit of the vineyards and olive groves that you did not plant" (Joshua 24:13).[32] Acosta's citation of biblical authority inadvertently undermines his treatment of profitable violence as a form of labor, for Joshua's admonition to his fellow warriors to remember the source of land does not originate in autochthony. Rather, Joshua stresses the distinction between the belonging that comes from labor—toil in the fields—and the possession of land through military conquest.

At the same time, though the text in Joshua could imply that the Israelites might not ordinarily be worthy of the land they now hold, they *do* hold it, *legitimately*, because of their covenant with God. The passage thus prefigures the Schmittian conception of the sovereign as the decider of the exception, and it is to this arbitrary adjudication that Acosta ultimately submits the matter of the conquistadors' compensation. And yet, as I signaled earlier, the sense of belonging that comes from burial is a thorny one, and even in the Book of Joshua it is not resolved by military conquest. In the last verses of Joshua, as Acosta must have known, Joseph's remains are brought from Egypt and buried in Shechem on the tract of land that Jacob—Joseph's father—had purchased for one hundred pieces of silver generations before the military conquest spearheaded by Joshua. Thus the parallels Acosta makes between Israelite conquest prior to the New Covenant and the Spanish Conquest of the Indies—the Christian interpretation of Joshua as *prefigura* for Jesus is well established—do not resolve the legality of conquistadors' grave looting, even as Acosta raises the ghost of the Indigenous dead through invocations of biblical authority on sovereignty in matters of (promised) land distribution in the spoils of war.

In addition to questioning the legality of burial spoliation, Las Casas raises the matter of Indigenous knowledge about grave sites. As seen in the passage of *De Thesauris* cited above, Las Casas asks how this treasure, which can be found only through the knowledge of native inhabitants, can be the object of conquista: "The question is whether all this [treasure] shall belong, indiscriminately, to whomever that, or by his authority, or with license by our Spanish monarchs, or by the governers who in their name head the governments of these regions, search [*quesierit*]." Here, his use of the verb *quaerere* brings to the fore the relationship between inquiry into Indigenous knowledge practices and possession through profitable violence (*conquest*), an etymology recalled by Gibson (1977) and discussed in the first two chapters above.[33] Las Casas then begins his own inquest into autochthonous wealth by recalling similar practices of burial among all peoples of antiquity, a narrative tactic he had already employed in the *Apologética*. For Las Casas, comparison to the burial practices of antiquity, both pagan and Abrahamic, allowed the conclusion that the original intent and honors of all burials could only be undone, willingly, by the owners of the graves or their descendants. Foreign kings could not even inquire, that is, discover, *con-quaerare*, their locations and belongings. Since the moment of the first discovery, by Columbus, most of the Indigenous peoples of the Americas had not proferred the wealth of their ancestors to the conquistadors of their own free will. In consequence, the crown and the conquistadors remained indebted to them and their descendants.

Moreover, Las Casas suggests that the entire conquest was suspect, as the wealth of burials had served to fund subsequent expeditions (i.e., its scalability in the modus operandi of venture capital through contiguity). Furthermore, the Papal Donation had given the Spanish monarchs the responsibility to evangelize in the Indies but not that of having dominion over them. Responsibility, for Las Casas, included providing the material means for missionary work. Yet even as late as the 1560s, most capitulaciones did not provide the salaries for priests associated with the ventures; instead, Indigenous subjects were expected to tithe, in addition to providing tribute, in labor and in kind, to the encomenderos or *corregidores* (chief magistrates or mayors appointed by the crown). For Las Casas and other Dominican friars, particularly in Peru, this situation was untenable, especially when imperial expansion had depended on priests for labor and, in some cases, capital. Priests—as Las Casas knew all too well from his own early experience as a priest affiliated with conquests in Hispaniola and Cuba before he joined the Dominican order—had also participated in the spoliation of Indigenous tombs. As I mentioned in the introduction, a notorious example given by Las Casas is that of Hernando de Luque for his founding role in the Compañía de Levante. Hernando de Luque had participated in the conquest of Peru through his partnership with Francisco Pizarro, Diego Almagro, and the licenciado Espinoza and had profited from the huacas' spoliation even as he did not physically participate in the looting of graves. The well-documented sources of wealth used for and reaped by the conquest of Peru allowed confessors working in the vein of Las Casas to explore in detail corporate and individual liabilities.

In Lima, the mendicant orders led by Archbishop Loaysa heralded the decade of the 1560s with *Avisos breves para todos los confesores del Peru acerca de las cosas que en él suele haver de más peligro y dificultad* (1560). Based on Las Casas's earlier *Confesionario* (1552), the twenty-six articles of the *Avisos breves* created a manual for confessors that provided the counterpart to the decision-tree-like manuals of the capitulaciones for venture capitalists. Yet the Peruvian *Avisos breves* took the Lascasian *Confesionario* one step further: they tried to account for all parties involved in the business of conquest, weapons dealers, merchants, and servants. Essentially anyone who had received something in specie or in kind from a conquistador was in danger of not receiving absolution from confessors who subscribed to the *Avisos breves*. According to the *Avisos breves*, those Christians who had any doubt about the source of their usufruct had to provide restitution to the Indians. Thus the place of doubt in the Catholic economy of grace and absolution became increasingly salient as the mendicants' examinations of conscience traced the ramifications of partnerships in the conquest of Peru, identifying who had profited or continued to profit from primitive accumulation.

In this context of de facto excommunication for an entire class of Catholic subjects, Las Casas most likely penned the *Tratado de las doce dudas* in response to the twelve doubts put forward by a fellow Dominican friar, fray Bartolomé Vega, who consulted him as a moral authority (Denglos 1992, xvi–xxxv). Although the polemics of the Valladolid debate (1551–52) on the merits of the conquest's legality have rightfully captured the imaginations of scholars of the early modern period and international law, it was the unique place of doubt, rather than polemic, that informed the experience of conquest and the understanding of self for all parties involved in the Spanish Empire's political theology, which was also an *economy* of grace.

Doubt in Circulation

Doubt for a Catholic subject, as a practical matter, is a mortal sin. As opposed to a debate on the state of a question, which can be discussed on the merits, a doubt has to be resolved a priori before a Catholic subject may act in good conscience. Moreover, performing an action with a doubtful conscience without justifying the act beforehand could prove even more problematic when the subject in question is a Catholic sovereign. As seen in chapter 2, the Laws of the Indies are riddled with the sovereign's doubts. Ultimately, Las Casas's *Doce dudas* touched on the doubt of a decider, the sovereign, and on the lawfulness or unlawfulness of an action, that is, the law itself. Las Casas would not free the deciders of the exception—Schmitt's definition of the sovereign three centuries later—from their doubts on the exception. Doubt, which had been a matter of moral theology, was treated by Las Casas as political theology avant la lettre.

In *De Thesauris*, Las Casas had implied that the sovereign could not be absolved of mortal sin unless he or she decided in favor of the liberty of the Indigenous people, whose violation meant that the Spanish monarch and his or her subjects had to make contrition in order to be absolved of past sins. However, according to the Council of Trent (1545–63), to achieve true contrition, the penitent had to resolve to sin no more, especially to not commit the *same* sin. Thus doubt, sinfulness, and the Laws of the Indies were inextricably linked in the moral arc of the conquest. In the *Doce dudas*, the doubts of the sovereign, which had been expressly manifested in the preambles to the Laws of the Indies promulgated in 1526 and 1542, required an urgent resolution. The economy of conscience, the weight and measure of it, so famously articulated by Charles I in the *Ordenanzas* of 1526 and later the New Laws (1542), were for Las Casas a manifestation of a sovereign conscience in a state of sin.

From Las Casas's treatment of the Christian sovereign's doubts, it can be inferred that the reiterated promulgation of laws in a state of ambivalence was sinful, but enacting laws that codified the ambivalence between caritas and cupiditas was doubly pernicious. Contrition, as codified at Trent, was either understood as an act made out of love for God (perfect contrition) or one made out of fear of eternal damnation (imperfect contrition) but in either case performed with the clear intention *not* to repeat the sinful act. However, the repeated pronouncement of doubt coupled with the redefinition of Christian love to coincide—at times—with cupidity stretched out the temporality of the conquest, with all its sins, especially *scandalous*, so that it was ongoing, self-perpetuating, with no *after* on the horizon. As his last will and testament, Las Casas willed his sovereign to recognize that caritas and cupiditas were antithetical. It was as if Las Casas, ever ironic, were asking of his sovereign, *you must decide* or, rather, *you ought to have decided* long ago.

The thorny question of the sovereign's liability touched directly on the recourse to exceptions in the venture capital model and their role in the denouement of the state of exception in the conquest of America. The sources for the first shipments of bullion from Peru were well known: ceremonial centers had been sacked as payment for the ransoms of Indigenous leaders such as, most famously, Atahualpa, but also the curacas of Lima, Pachacamac, Jauja, and Huancavelica. The objects made of gold and silver, feathers, paint, stones, and ceramic had been stripped down to their metal base and melted into bullion in Cajamarca before they were distributed, packed, and shipped to Panama and then Seville. Investors and partners, such as the licenciado Espinoza, had to be repaid in Nicaragua and Panama. The king had to receive his carried interest, or quinta real. By dedicating an entire treatise to the lost treasures of Peru *and* the illegitimacy of the conquest as a whole, Las Casas exploited the extensive documentation of grave looting in Peru to expose the insurmountable debt of the Spanish Empire, which had been accumulating material obligations to the *peoples* of the Indies from its very inception as a series of venture capital funds.

Incurring mortal sins for crimes against jus gentium, which Las Casas argued were crimes against God's covenant with Christians, *and* amassing debt to the Indigenous dead and their descendants placed the very conscience of the sovereign in hock to his new subjects: how could it all be repaid, let alone absolved? Las Casas refused to privilege the translation of the Indigenous *sacer* into the metropole's *capital.* In doing so, he upheld the alterity of the huaca present in the *castellanos de oro,* or pieces of eight, that were changing hands in European money markets. To the credit of Las Casas, he raised the specter of the Indigenous ancestors over the credit of the Spanish sovereign.

An object of wonder to the Portuguese ambassador, who was received by Philip II in his Audiencia, the credit (*crédito*) of the Spanish sovereign was indeed formidable, as attested by a chest full of bullion on display in his throne room (Sanz Ayán 2004, 21). Yet this ostentatious wealth belied a constant lack of liquidity. In 1557, for the first time, the Spanish crown had to default on its debt service to its creditors, marking a shift in the crown's way of financing the state and its wars on the European continent. The experience with bankruptcy had led Philip II, the Council of Castille, and officials of the House of Contracts to ask the encomenderos of Peru and the new viceroy, the Conde de Nieva, to make him an offer for concession of the encomiendas in perpetuity. Frays Domingo de Santo Tomás and Bartolomé de Las Casas famously responded on behalf of the curacas of Peru by making Philip II a counteroffer to buy their self-determination in pesos of gold and silver, the subject of the next chapter.

Whereas the crown's quinta from its American ventures had figured prominently as a source of credit before the 1557 default, the full haul of the treasure galleons increasingly underwrote loans for the crown after that first crisis of faith in Philip II's credit. And yet the ramifications of *De Thesauris* for the empire and its financiers after 1557 were all too clear. Servicing one kind of obligation (financial) while ignoring another (spiritual) made the good faith of these enterprises entirely dependent on the worst kind of theft, divine and profane: of the time for salvation and of the wealth of the Indigenous dead and their descendants. Almost immediately following passage of the *Ordenanzas* of 1573 and almost ten years after the death of Las Casas, the crown would once again find itself unable to fulfill its obligations, material or otherwise.

Yet, for Acosta, the cupiditas that fueled the Spanish Conquest was a godsend, for it brought Christianity to dominions that had been ruled by the devil.[34] Not so for Las Casas, who had contended in *De unico vocationis modo* that the devil and idolatry entered the Indies only with the greed of the conquistadors, not before. Acosta, then, condoned grave robbery in the name of saving the Indies from the scourge of the devil in his subterranean domains, a residence in the Indies that, according to him, the devil had claimed *before* the conquest. Therefore, in Acosta's view, the looting of graves by conquistadors, though unsavory, could be interpreted as doing God's work since the spoliation also performed the extirpation of huacas and their accompanying narrative objects, the khipus. Recalling his insistence on treating profitable violence as labor, the de facto extirpation of idolatry and the devil's residence had to be compensated in some way. It was, after all, hard work, and therefore it needed compensation. Coupled with the promotion of the material comforts of "civilized life" that had been brought to the Indies by enterprising conquistadors, their grave looting was not scandalous, or even a necessary evil. It was a Christian act.

As discussed in chapter 2, by explicitly outlining the paternalistic goals of civilizing the Indians so that they could lead a "civilized life," the *Ordenanzas* of 1573 fully endorsed the Aristotelian civilized/barbaric binary that had been promoted vociferously by Sepúlveda. Acosta formally aligned himself with Sepúlveda's Aristotelian arguments, even as the Jesuit claimed disagreement with Sepúlveda on the just causes of war. "Ut enim Barbari" (The Barbarians), according to Acosta,

> veluti mixta humana & ferina natura constant, ut moribus non tam homines, quam hominum monstra videa[n]tur sic quae cum illis instituenda est consuetudo, partim humana & liberalis, partim subhorrida, & ferox sit, necesse est, usq[ue]; dum nativa illa sua feritate deposita, paulatim mansuescere incipia[n]t, & ad disciplinam humanitatemq[ue]; traduci. (Acosta 1589, 254–55)

> [display a nature that appears to be a mixture of the wild animal and the human being, and their customs are such that they appear to be human monsters [*hominum monstra*] rather than humans as such. So, we must begin a treatment of them that is partly human and partly animal, until they begin, little by little, to lay down their native wildness, and become docile and accustom themselves to discipline in proper human customs.] (Acosta 1996, 1:82)

The imposition of sedentary life; the *reducciones* (settlements of Christianized Indians); the cultivation of wheat, grapes for wine, and olives for oil; the wearing of shoes—all had been benefits of "policía" and Christianity as outlined in the *Ordenanzas* of 1573. Under Viceroy Toledo's reforms in Peru, Acosta was familiar with their implementation.[35] In law 141, commonplace figures for Indigenous monstrosity and savagery go hand in hand with a litany of objects that characterize the good life brought by the Spanish.

Y los tenemos en paz para que no se maten ny coman ni sacrifi-
quen como en algunas partes se hazia y puedan andar seguros por
todos los caminos tratar y contratar y comerçiar aseles ensenado
puliçia visten y calçan y tienen otros muchos bienes que antes les
heran prohibidos aseles quitado las cargas y serbidumbres aseles dado vso de pan vino azeyte y otros muchos mantenimientos
paño seda lienço cauallos ganados herramientas armas . . . y que
de todos estos bienes goçaran los que vinieren a conoçimiento de
nuestra santa fee catholica y a nuestra obediencia. (Morales Padrón 2008, 516)

[And we have pacified them so that they do not kill or eat or sacrifice one another as they did in the past. And they can travel and
trade and do business on the roads, safely; we have taught them to
live in polite society. They dress and wear shoes and have other
goods that were previously prohibited to them. We have removed
their burdens and servitude and given them the custom of [eating
and drinking] bread, wine and oil and other sustenances. Cloth,
linen, horses, livestock, tools and weapons . . . all these benefits
they will enjoy if they come to the knowledge of our holy Catholic
faith and to obey us.]

Like Toledo's push to use native informants to delegitimize the Incas
of Peru and, thus, construe the Spanish as liberators of native tyranny,
the law combines a narrative of past grievances, such as monstrosity
(cannibalism) and servitude, with an enumeration of material and spiritual benefits, which include an enforced "peace" that is conducive to
free movement for trade. Obeying the law, becoming a subject, then,
entailed a process of self-effacement and scripting: first, accepting the
imputed monstrosity; then, reifying the practices and traits associated
with it; later, rejecting them; and, finally, learning to value the temporal
and spiritual "benefits" of the new regime. Becoming "civilized" included evaluating self-interest—temporal and spiritual—in this new
paradigm. Similarly, Acosta's calls to "tame monstrosity" echo the
sentiment and letter of the law that promises benefits and threatens
punishment to ferocious men. At the same time, Acosta subscribes to
Vitoria's analysis of the just causes of war so that his justification for

the Spanish use of force in the Indies remains a "defense" of free movement for trade and evangelization, whereas dominion is explored in
Aristotelian terms.

In the first and second books of *De procuranda*, Acosta recalls the
arguments of Vitoria, Sepúlveda, and Las Casas to his readers. He sum-
marizes and approves of Vitoria's reasoning thus:

> Praeter hanc causam acceptae iniuriae, aut violati iuris gentium,
> nullum nostri maiores iustam agnouerunt, neq[ue]; gloria quaeren-
> dae, neq[ue]; cumulandarum opum, neq[ue]; amplificandi domina-
> tus, neque verò religionis propagandae. Quotquot verò no[m]laesi
> arma sumpserunt, eos praedonis potius, quam milites vocitandos
> censuerunt. (Acosta 1589, 222)

> [Without this cause of harm received or the violation of human
> rights, our peers did not recognize any other form that was deemed
> a *just cause*. A *just cause* cannot be for gaining honor, nor for the
> accumulation of wealth, nor for extending dominion, not even that
> of propagating holy religion. Those who, not having received any
> type of harm yet took up arms, were judged to be worthier of the
> name bandits than of soldiers.] (Acosta 1996, 1:64)

Yet Acosta will argue vociferously that the conquistadors are not in fact
bandits, even though their unbridled greed may have given cause for
such an imputation. Moreover, he treats Spanish dominion as a fait
accompli and accuses Las Casas, though never by name, of pitting temporal and spiritual powers against each other.

Acosta condemns the unbridled greed of the conquistadors and the
missionaries, who have succumbed to the corrupting influence of
riches in the Indies (*De procuranda* I.xi). Self-criticism, Acosta argues,
would recognize that "plusquè operae in colligendo argento ponamus,
quam in acquirendo populo Dei" (we are more involved in money-
making than gaining people for God) (1589, 180; 1996, 1:37). This
self-serving attitude, he says, has led the barbarians, the "barbari," to
conclude, understandably, "ne vaenale pute[n]t barbari esse Eva[n]-
geliu[m,] venalia Sacramenta, neq[ue]; animas nobis curae esse sed
nummu[m]" (that we charge for the Gospel and the sacraments and

that we only care for the money and not for their souls) (1589, 179; 1996, 1:37). Instead of self-interest and the pursuit of profits, missionaries and explorers should succumb to love for the barbarians as the motivating force behind evangelization (Acosta 1589, 160–66). Indeed, Acosta, citing Chrysostom, argues that the rewards for converting a soul are much greater than riches (141–47). Then, following the examples of the Gospels and patristic authorities, Acosta likens souls to riches. Thus it is only when self-interest is wed to love for the other that Acosta feels that he can properly speak of sanctified cupiditas without contradiction.

At the same time, although Acosta denies acting out of self-interest in the pursuit of temporal riches, he admires entrepreneurs and reiterates the material benefits—the civilizing of *hominum monstra*—that the Indians enjoy as subjects of the Spanish crown. In this way, entrepreneurial zeal provides a model for missionary work and offers the material benefits for Indigenous conversion and civilization but, in excess, corrupts and is materially and spiritually inefficient. Yet the effects of unbridled greed have been to strip the Indies of their "ancient prosperity" and to diminish the Indian population (Acosta 1589, 183–85). Acosta then levels his greatest moral attacks not against the conquistadors but against Las Casas and his adherents. In his discussion of the proper means for conversion, he recognizes that the "gospel of peace" and the "sword of war" joined together are a paradox, but it is love and, perhaps, a leap of the imagination that are able to unite them even when the understanding cannot (208–11). In this way, though he subscribes fully to Vitoria's argument on the just causes of war and titles of dominion, he will deny the salience of peaceful evangelization that Las Casas outlined in *De unico vocationis modo* taking the first apostles as his model.

For Acosta, the intellectual rigors of the Greeks and Romans posed a formidable challenge to Christ's humble apostles, rendering miracles a necessity to overcome disbelief and instill faith (1589, 247–50). In an implicit refutation of the combined arguments of Las Casas's *Apologética historia sumaria* and *De unico vocationis modo*, Acosta then proceeds to reject evangelization in the manner of the apostles by relating the following three premises: first, Indians are inferior to the ancient pagans in culture and rational thought; second, God limits the performance of miracles to peoples and places where they are most needed; and third, potential martyrs should follow God's lead in treating their own lives as a limited resource, similar to miracles. Thus, for Acosta, peaceful evangelization and its associated martyrdom are wasteful in the West Indies because they do not make discerning use of missionaries' lives, as Indians are more inspired to belief than their Greek and Roman counterparts, but their "ungodly" habits are more difficult to extirpate. Acosta rejects the argument that attributes the dearth of miracles in evangelization campaigns to the missionaries' impurity. Instead, he interprets the circulation of grace only as a function of Indigenous worthiness to receive it, bearing some resemblance to the validation mechanisms associated with the merchant's crédito. What follows is a narrative of wasted charity.

Acosta's examples of "wasteful" martyrdom include the deaths of priests in La Florida (1589, 247–50). He does a cost-benefit analysis of martyrdom in greater detail, condemning what he perceives to be the self-interest of martyrdom seekers. According to Acosta, these missionaries showed a lack of prudence for "specie iustitiae sanctioris committere, ut salute ipse abijcias, & alienam nihilo amplius compares" (under the guise of greater holiness [they nonetheless] gain no lives from it) (1589, 238; 1996, 1:73). Indeed, nothing was more wasteful than becoming the object of monstrous anthropophagi, which would not have resulted in "true martyrdom" in any case. What, then, constituted a "true" as opposed to a "false" martyrdom? Acosta's rejection of martyrdom in the Americas could not have been expressed in more visceral terms.

> Neque vero ab istis Martyrium expectandum est, quae fortassis spes tantum discrimen levaret, non enim pro Fide, pro Christo, pro Religione moriendum est: Sed ut vel Suauiores epulas de te praebeas, quod Brasiliensibus, & toti Septentrionali orae huius orbis vulgare est, vel spoliu[m] praebeas barbaris elegans, vel deniq[ue]; quia visus es nu[m]quam, & quid in te sibi liceat, experiri iuuat. (Acosta 1589, 239)

> [And it is not as if a true martyrdom would await us, which would be a great relief in such labors. For death would not come to us for

the Faith, for Christ, or for our religion, but rather to make us a more succulent morsel for their palate, as it is common in Brazil and in all the northern coasts of the New World, or to become a hunting-trophy for them, or, finally due to the fact that they had not seen a foreigner to experiment and see what they could do to us.] (Acosta 1996, 1:73)

Let us put aside the objections of the missionaries who died in precisely the manner derided by Acosta and were later canonized for their martyrdom by the church. Acosta's reasoning showed an almost utilitarian disdain for the deaths of his brethren in Christ, *unless* they could have maximized (i.e., scaled) the number of neophytes achieved per death. Yet Acosta not only frowned on martyrdom in the Indies in quantitative terms. He analyzed the efficiency of martyrdom based on his perception of the lesser worth of Indigenous neophytes' souls. In the narrative supplied by *De procuranda*, miracles and martyrdom were wasted on the neophytes of the West Indies; they were not worthy of *charis*.

Having refuted the arguments for peaceful evangelization with his cost-benefit analysis, Acosta proceeds to analyze another two options for evangelization: to preach among those who "are already subjected justly or unjustly to the Christian princes" (1589, 240) or to seek out new converts among those who have remained beyond contact, with a garrison of soldiers for an escort. Acosta commends both options. The greatest confusion in his thought emerges precisely when he attempts to dispel doubts on the viability of conversion among those "already subjected justly or unjustly," for Las Casas had explored such doubts to the ultimate logical consequences in the *Doce dudas* (1564).

Acosta pursues Las Casas with a relentless circular logic where prudent evangelization (i.e. under the protection of the state) is the ultimate telos. Thus, the crown and the church must uphold their partnership, even if *past* actions were scandalous. We saw in chapter 1 how venture capital engages its partners in a metalepsis of the future; after 1573, the same can be said for Acosta's attitude to the past. Acosta laments that Las Casas could have promoted such a schism between the crown and the church in order to rectify the scandals of the conquest that were buried in the past. Acosta treats the conquest like a recent antiquity; narratively, he passes it off as a recent antiquity, similar to the *Requerimiento*'s treatment of the Papal Donation, as discussed in chapter 2. Given that the conquest was a thing of the past, it had been reckless and inefficient of Las Casas, Loaysa, and others to demand the restoration of Peru, and all of the Indies under Spanish imperium, to their native lords. Himself a Spanish subject, Acosta may have felt the weight of Las Casas's concerns about the mortal sin of the Spanish sovereigns and their subjects unless they abandoned the Indies. Yet these concerns were weighed on a merchant's scale, much like Charles V's burdens of conscience, which, for Acosta, had long been forgotten.

Acosta's arguments, doubts no longer, sought shelter in the fait accompli of the conquest and the state's constituted powers—de facto or de jure—to protect the already converted (one of Vitoria's just titles of war) at any cost.

> Nihil perinde instructioni, & saluti Indorum nocere compertum est, atq[ue]; perversam qua[n]dam, & malignam potestatum temporalis, & spiritualis concertationem, aut imminutionem, aut quoquomodo offensionem. Atq[ue]; ut de caeteris modo magistratibus saecularibus taceamus, certè grauiter erra[n]t quidam specie fortasse pietatis, ius regiu[m], & administrationem vocantes in dubium, quaerentes interdum, quo titulo, & iure Hispani domine[n]tur Indis? Num haereditario iure ad nos deuoluti sint, an bello iusto subiecti? (Acosta 1589, 251)

> ――――――

> [It is a well-known fact that nothing causes such damage in the instruction and the spiritual welfare of the Indians than the competition between the two powers, i.e. the temporal and spiritual, and the deterioration of relationships or any other type of struggle with the civil powers. Let us leave aside for the moment other types of secular magistrates, and focus on those who, under the guise of piety, cast doubt upon the right of our kings and their government and administration. They are, without a shadow of a doubt, making a mistake. They are the ones who stir up disputes over the rights and terms through which the Spaniards dominate the Indians, and the question of whether Indians are just an inheritance transferred

from their princes to ours, or if we have merely conquered them through a just war.] (Acosta 1996, 1:80)

Acosta then justified his refusal to pursue the logical consequences of reasoning on the manner, means, and time of conversion with a new-found expediency. Even though the crown had sought to avoid the conquest at all costs, by insisting on the "new methods" for exploration, discovery, and pacification in 1573, Acosta conceded the legitimacy of its absolute command, *merum imperium*, with little thought about restitution. This would imply that Acosta felt there was no moral risk for preachers involved in forced conversions, in illegal encomiendas, or in a continued partnership with an unjust temporal power. Spiritual profits for the church trumped any other consideration.

Quae sanè disputatio eò pertinet, ut administrationis Indicae, vel tollatur, vel certè debilitetur autoritas, quo semel si gradus fiat, qua[n]ta sit futura pernicies, quae perturbatiorerum omniu[m] co[n]sequatur, dicivix potest. Neq[ue]; verò id ego modò suscipia[m], ut bella, belloru[m]q[ue]; gestorum rationes defendam, atque omnes illos superiorum temporum turbines. Illud religiose & utiliter moneo non oportere in hac causa amplius disceptare, sed veluti praescriptum iam sit, optima fide agere debere Christi Servum. (Acosta 1589, 251)

[These sorts of polemics might lead us to abandon the dominion or the administration of the Indians, or, at the very least, greatly debilitate our control over them. And if we begin to yield to these sorts of opinions, and we do not reprimand them with a firm hand, we cannot say what sort of evils and universal ruin may follow, and the grievous perturbation and disorder in everything. Now it is not that I am proposing here to defend the wars of the past and all that happened through them nor all the revolutions and disturbances that have taken place, but I am warning as a supremely useful religious piece of advice that *it is not worthwhile* going on arguing over this matter, but rather the servant of Christ in good faith should take it as a fait accompli.] (Acosta 1996, 1:80; emphasis mine)

Writing almost two decades after the deaths of the Incas of Vilcabamba, Titu Cussi Yupanqui, and Tupac Amaru, Acosta did not deny the possibility of a return to original rule but rather argued against its *value*: in time, in labor, and in (good) conscience. Like the contingency plans made law in the *Ordenanzas* of 1573, what Acosta offers his brethren in Christ are ramifications for their (im)proper conduct and speech in the Indies. The pursuit of truth and justice, a sovereign's examination of conscience, or that of his subjects, these are all *happenstance* for insurgency. The risk of insurgency and apostasy was unacceptable among the gamut of risks that could or even ought to be taken, though merchants and conquistadors should be commended for risking their lives and livelihoods in pursuit of profits, notwithstanding any doubts they may have had.

Without a doubt, it is ironic that a Jesuit, professed in the tradition of Loyola's *Spiritual Exercises*, would advocate, wholeheartedly, against examinations of conscience. Or, to be more precise, Acosta relegated the matter of conquest, love, and interest to a "polemic" (*disputatio*), purging it of the urgency of the *dubium* that, for Las Casas and his followers, had posed the existential threat to all those who had benefited from the collective enterprise in profitable violence known as conquista. To this day, the questions posed by Las Casas haunt us still. *Who* are the true Christians? *What* is a true Christian?

In Acosta's critique of conquest and prescriptions for imperial reform, Philip II could once again indulge in the doubt of a sovereign, though the discourse of polemic would reify it, allowing the sovereign to emerge from the impasse. Acosta was critical of what in his view the conquest had been, but his solutions were necessarily logical extensions of that conquest's practices. The conquest's logic had already become second nature to its most ardent critics, such as Acosta, and even, as we saw earlier in this chapter, to Vitoria. However, the memory of the Lascasian treatment of the sovereign's doubt and Indigenous freedom remained alive and well in the Viceroyalty of Peru. The next chapter is dedicated to the Indigenous reception of the ambivalence inherent to the enterprise of conquest and begins with the counteroffer made to Philip II in 1560 by the curacas of Peru for dominion over their native lands.

CHAPTER FIVE

The Bidding of Empire

The Curacas Negotiate Dominion with Philip II

It was no horse's head, but in 1560, the native elites of Peru made Philip II an offer that must have been difficult for the Spanish monarch to refuse.

> En nombre de los indios del Peru, contra la perpetuidad; y ofrecen servir con lo mismo que los españoles y cien mil ducados más; y si no hobiere comparación de lo de los españoles, servirán con dos millones, pagados en cuatro años, con las condiciones que ponen. (Las Casas and Santo Tomás 1957, 465)
>
> ───────────
>
> [In the name of the Indians of Peru and against perpetuity, they offer to see the Spanish [conquistadors'] offer and raise it by 100,000 ducats more; and if the Spanish make no offer, then they will give two million [ducats], paid in four yearly installments, with the following conditions.]

By vowing to see and raise one hundred thousand ducats more than any offer Spanish conquistadors in Peru had made in the past or might make in the future, the native elites of Peru instigated a bidding war for the right to full control over Indigenous labor and lands in Peru. Yet as they bid more than two million ducats in Andean bullion to purchase something from King Philip II of Spain, they would also argue that it was not his possession to give or take away: namely, sovereignty over the lives, livelihoods, and lands of the Indians of Peru. In addition to making a bid to purchase something they ostensibly already had in their possession, the curacas—native elites in Quechua and Aymara— incorporated another contradiction in their monetary and moral appeals to the king. They couched their capital investment in this new *sociedad*—business partnership but also society—as an effort to revive past practices of community building in the Andes. A paradox thus emerged from this negotiation: in seeking to revive the Andean past in order to practice it, its proponents treated this productive past as an inert past, a commodity for trade with the king of Spain. It was inalienable but also up for grabs.

Peru offered a dynamic, constantly changing landscape of institutions and alliances in the 1560s. Manco Inca's insurgency in 1536 and later retreat to Vilcabamba was followed by the wars among various Spanish factions as the original partnership between Francisco Pizarro and his brothers and Diego Almagro fell apart over who would control Cuzco, center of the Inca empire. Local Indigenous elites (curacas) and Indians (indios) cast their lots in these conflicts, which were rending at the seams the original corporation behind the conquest of Peru. In this context, the Indians held in encomiendas before the First Lima Council of 1551, felt little Spanish influence in their daily lives except for those areas of intense interaction such as Cuzco and Lima, areas with preconquest accumulations of wealth in labor and precious metals (Lockhart 1994, 28–30). Despite the desire of the encomenderos to live near the indios, who had been placed in their care for spiritual tutelage in exchange for their labor, the only non-Indigenous subjects allowed to do this were the *doctrineros*, those who were sent to live among Indian neophytes and teach them Christian ways.

It has been argued that the encomienda system in the Andes was grafted onto the existing institutions of *ayllu*, *minka*, and *mita*, which native elites had managed as tribute to the Inca hierarchy. Curacas were prominent members of a large kin group (ayllu) that claimed a huaca as a common ancestor; the huacas, the ancestors embodied in mummies (*mallki*) and uncanny land and water formations, spoke to their descendants, involving themselves in the lives of those who venerated them. Curacas had always found themselves in tension between the demands that their kin and huacas made on them at the local level and the requirements of pan-ethnic overlords such as the Incas at the supralocal level.[1] Minka was and remains a reciprocal labor exchange within the ayllu, and it is given priority over labor tribute to the state (mita). As numerous authors have contended, curacas had access to the excedents of the ayllu, both in labor and in kind, through their control of women and the practice of polygyny and goods produced by women (e.g., corn beer and textiles).[2] The conquistadors recognized the importance of alliances with the curacas, but the encomiendas were not distributed in accordance with the traditional boundaries of the pre-Hispanic ayllus. The territories covered by encomiendas were not necessarily contiguous, creating a patchwork of competing encomiendas that were also at odds with the territorial delimitations of the ayllus. This discrepancy in the organization of kin, territory, and labor created upheaval in traditional power structures. It upended not only the hierarchy among greater and lesser curacas but also the animated landscape of power among the mummies and huacas that continued to speak to their kin but could no longer perform the organization of ayllu, minka, and mita coherently.

The mismatch between institutions in the Andes, the desolation of the indios, and the increasing campaigns to extirpate idolatry led to the huacas abandoning their traditional vessels. In the 1560s, followers of the *taquiy ongoy* (dancing and singing sickness) movement sought to revive the huacas by providing themselves as places where the huacas could recover their strength, enough to fight against the Christian God and saints and expel the Spanish once and for all. The *taquiongos* (practitioners of taquiy ongoy) demanded that the curacas purify themselves in order to communicate with the huacas once more; these demands, in turn, implied that the curacas had been "contaminated" by their contact with the Spanish and, thus, lost their ability to perform the will of the huacas. As Stern (1993) has contended, the ambivalent attitude of the curacas to the demands of the taquiongos exemplified the conflicts of interest experienced by native Andean elites. However, the curacas' bid for incorporation with the crown in 1560 represents a period, if only for a couple of years, when the interests of encomenderos and curacas did not align. What is striking is that both parties would claim to genuinely represent the interests of the Indians.

The system of labor tribute in exchange for Christian tutelage, or encomienda, relied on the pre-Hispanic habitus of labor and kin relations between the curacas and the ayllus, all the while upending its traditional boundaries. The tribute of the indios not only provided material rewards to the encomenderos and the royal coffers but also paid for the salaries of the doctrineros. The various political visions that emerged in Peru during the decade of the 1560s all had to address this foundational contradiction in the business of Indian labor in exchange for spiritual exchange. The alternative proposed by the curacas in their bid for incorporation with the Spanish crown deserves critical attention because the curacas were widely seen, by all parties, as the mediators of the spiritual and material economies of the empire.[3] Yet their mediation between ayllu and empire, at the time of the competitive bidding for perpetuity of the encomienda or incorporation, remains difficult to define.

What does it mean to *negotiate*? In the edited volume on the native peoples of New Spain and their negotiation of the terms of their submission, Kellogg (2010) uses the phrase "negotiation within domination" to refer to the participation of Indigenous elites in the development of colonial institutions. Yet this focus on Indigenous elites could serve as another instance of a system predicated on indirect rule, whereby the metropole relies on Indigenous elites to organize domination of productive classes in the service of capital. How, then, might we describe the alliances between encomenderos and Indigenous elites *against* the metropole? As Ruíz Medrano (2010, 47) has argued for sixteenth-century New Spain, the last direct descendant of the Mexica ruling class, don Luis de Santa María Cipac, who was also the governor of México Tenochtitlán, allied himself with the Spanish encomenderos in a revolt against Philip II that would have led to the establishment of a new monarchy with Martín Cortés, the second marqués del Valle, at its head. The encomendero and Mexica revolt responded to changes in law that redirected most tribute to the king, leaving little for local elites.

Could the participation of Mexica nobles in the encomendero insurgency in the city of Mexico (1564–66) be qualified as an instance of *Indian* political consciousness, as contended by Ruíz Medrano?[4] Alternatively, it could be construed as a bid by Indigenous elites for increased power in a new empire through a self-fashioning of authenticity as representatives of Indian sentiment. Unless *Indian* consent to Indigenous elites is taken as a given, even as the nature of that consent must be questioned, the synonymous use of "Indian" and "Indigenous elite" proves problematic for our understanding of the power dynamics in the Spanish Empire. Indian interest could very well be yet another trope from the rhetorical arsenal of Indigenous elites wielded to gain legitimacy against Spanish elites. As Guha has argued in both *Elementary Aspects of Peasant Insurgency* (1999) and *Dominance without Hegemony* (1997), colonial structures of power need the participation of Indigenous elites in order to maintain productivity. Perhaps the synonymous use of "indio," "curaca," and "cacique" ought to be avoided in discussions of colonial power structures in Latin America because the interplay of power, hegemony, and subaltern gets confused in the redaction. Guha's analysis of colonial history and power assumed significant distinctions among the Indigenous peoples of the Indian subcontinent as he privileged the consciousness of the subaltern in writing a history from below. The question remains, Did negotiation "counterbalance" domination, as suggested by Owensby (2010, xii)? Or is negotiation just another form of domination? Were Indigenous elites persuasive or coercive in their power relations with the indios on whose behalf they were ostensibly negotiating with the Spanish sovereign?

In their failed bids for incorporation and perpetuity, both the encomenderos and the curacas claimed to speak in the interest of the indios of the Andes, even as they engaged in parallel negotiations with the Spanish crown and never negotiated among themselves. Were the indios, indeed, their partners in these bids for new political organizations of tribute in labor and in kind? In the Andes, how did the general and limited partners in the Spanish Conquest confront the contradictions that came to the fore in these divergent proposals for doing the empire's bidding in 1560? While, as Kellogg (2010, 4) has contended, "it is also true that the ability of the Crown to assert authority—whether by Isabel in the Caribbean and early sixteenth-century New Spain or the Hapsburgs in later sixteenth-century New Spain and Peru—lay in part in the willingness of the Indigenous population to accept that authority," conducting negocios with another, in the busy-ness of it, demarcates liabilities and envisions a shared future as a function of mutual gain. Also, how is Indigenous *acceptance* of the crown's authority construed? Was it freely given? Coerced? The result of persuasive argumentation? As argued in chapters 2 and 4, the forms of Indigenous consent were under continuous review. Moreover, as seen in the previous chapter, the social contract(s) in the negocio of the Indies, according to Isabel's letter to Ovando, explicitly define the liberty of the colonial subject as a function of productivity. The distribution of labor, its usufruct, and its times of productivity and rest supersede any other concerns, including questions of justice. Indeed, the encomienda economy makes justice contingent on usufruct. This usufruct, in turn, depended on the Indigenous habitus, entirely at odds with the modus operandi of Christian empire.

This chapter examines the failed bids for perpetuity of the encomienda by the encomenderos and the counteroffer for incorporation by the curacas that ignored their erstwhile allies. The visions of a prominent dominican, fray Francisco de la Cruz, are analyzed in the final sections for their grand attempt to reconcile all contradictions between the Spanish modus operandi and the Andean habitus in the service of a new Christian empire centered in Lima as the seat of temporal and spiritual power. His advocacy for polygyny and the marriage of prelates within what he considered a coherent, very Catholic worldview ended with several trials by the Inquistion in Lima. Notably, the trials did not stop after he was burned at the stake as a heretic in 1578. Years later, his inquisitors could not decide if his attempts at reconciliation were a sign of madness (*locura*) or heresy (*herejía*). In turn, the binary opposition of heresy and madness in Peru in the late sixteenth century represents a clear break with Alfonso X's characterization of heresy as a form of madness in the *Siete Partidas* (7.26).

BANKRUPTCY AND BIDDING

As discussed in chapter 4, the crown's bankruptcy in 1557 lent new urgency to negotiations with the encomenderos and curacas of Peru. Following the bankruptcy, Philip II made the first overture to the encomenderos in the Viceroyalty of Peru, offering perpetuity of the encomienda but also greater jurisdictional power to the encomenderos in exchange for gold and silver. The terms Philip II offered to the encomenderos in 1559 surpassed the demands that Gonzalo Pizarro and his cohorts had made in 1546.

As presented by Philip II, the transition to perpetuity would have given the encomenderos limited sovereignty over native subjects but not over the territory they inhabited. Spanish residents with enco-

mienda (*vecinos encomendados*) could leave their labor and land trib-
utes to their inheritors, reside with the Indians, and have jurisdiction
over them. The monarch's proposed system never came to fruition. The
residency clause was particularly important to the encomenderos and
a cause of concern not only for the clergy and the curacas but also for
Spanish residents without encomienda (*vecinos sin encomienda*) such
as merchants and conquistadors or their descendants, who already
complained that the encomenderos had a monopoly on Indigenous
labor. For their spiritual care, the encomenderos had to pay a tenth of
the tribute to the clergy responsible for evangelizing among the Indians
who gave them tribute. This had led to many encomenderos having a
hand in assigning the doctrineros to preach in their encomienda; however, there were also many complaints leveled by doctrineros against
the encomenderos, claiming that they never received the tithes.

At the time of the negotiations for perpetuity or incorporation,
fewer than 500 of the 8,000 Spanish male residents in Peru were encomenderos; this handful of encomenderos received over a million
pesos of tribute per year from over half a million Indigenous men between the ages of eighteen and fifty. Though Indigenous women were
not accounted for as direct sources of labor, their bodies and the products of their labor provided the excedent among the ayllus and were the
foundation for the curacas' power and access to male labor for the encomienda. The other 7,500 Spanish residents were sons of conquistadors or *pacificadores* (peacemakers) who had not received encomiendas
but made claims to them. Then there were the vagrants, the *vagabundos*
(vagabonds) and *gente perdida* (lost folk), also known as *soldados* (soldiers), in short, the armed men who wandered Peru as mercenaries
looking to try their luck in the next insurgency. They had weapons but
had to beg for food and shelter. Many Spanish *pretendientes* (claimants)
to encomiendas feared that allowing the current encomenderos to purchase jurisdiction in perpetuity would cut short the slim chances they
might have had to receive an encomienda in the status quo.[5]

The encomenderos of Cuzco, La Plata, Lima, Trujillo, Chachapoyas, and Santiago de Moyobamba offered the king three and a half
million pesos to be paid in eight installments over eight years. This
must have been quite a disappointment to Philip II, who had entertained and rejected another offer for 7.6 million pesos from the encomenderos just six years earlier, when Antonio de Ribera advocated

on behalf of the encomendero community. Antonio de Ribera chose to negotiate with deference to the metalepsis of "love interest" whereby love for one's brethren not only did not contradict the interest one hoped to gain from one's brothers in Christ, but reinforced one's other interests. As proposed to Philip II, the encomenderos' love for the Indians would multiply just as the population would multiply, and the capital investment in land and people would similarly bear fruit. The qualms about unnatural metaphors, of money spawning money, were no longer the common currency of moral values, at least not for those members of the privileged social and economic station in Peruvian colonial life known as the encomenderos. As a sign of their good faith, the encomenderos accompanied their arguments for the perpetuity of the encomienda, couched in "love interest," with their estimation of the value, in pesos, of the right to jurisdiction and tribute over the Indians in perpetuity.

The arguments in favor of the encomienda in perpetuity in 1556 and 1561 did not emphasize remuneration for past actions. Three decades after the bidding wars, Acosta would give a much franker assessment of the capital investment and labor contributions of the conquistadors (when violence is accounted for as labor) and the crown's ensuing debt (and, to his mind, lack of relative power). Though for Acosta the sovereign's power in an empire would be proportional to the percentage of its original investment in *ultramar* (overseas) ventures, the encomenderos themselves had supplied a wider gamut of arguments to justify their rebellions and claims to the perpetuity of the encomiendas in the mid-sixteenth century.[6] Instead of speaking in terms of accounting for past debts, they made future projections and employed the metalepsis of venture capital, that is, the confusion of causes for effects. Notably, the encomenderos claimed that if they received their encomiendas in perpetuity they would no longer have a reason to rebel against the Spanish crown.

The circular logic of the encomenderos found support in the Franciscan Alonso de Castro, who presented a report in 1555 to Charles V and Prince Philip, at their request, on the subject of whether the encomiendas could in good conscience be sold in perpetuity. Castro sided with the encomenderos of Peru because of, *not in spite of*, their recent rebellion. For Castro, the Papal Donation could not strip Indigenous polities of their sovereignty, but it provided a "supra" sovereignty to bring knowledge of God, peace, and justice to the Indies. Had the encomenderos succeeded in their rebellion, Spain would have lost its supra sovereignty *in actu* and Christianity would have lost its privileged place in the Indies: the unfaithful would not convert, and the newly converted could become apostates. To postpone the granting of the encomiendas in perpetuity would only encourage more rebellions. Thus, in another nod to the metalepsis of venture capital, Castro recommended conceding the encomiendas in perpetuity in good conscience.[7] Not surprisingly, following Castro's recommendation and Antonio de Ribera's multimillion-peso offer, the Consejo de Indias and the crown, which was grappling with impending bankruptcy in 1556, sent commissioners to negotiate with the encomenderos of Peru.[8] However, the Consejo de Indias also issued a detailed *opinión* (judgment) that outlined all the drawbacks of perpetuity with jurisdiction, including de facto slavery in perpetuity for the Indians and more rebellions among the Spanish residents without encomienda (Konetzke 1953, 340–57).[9] The Consejo contended that the king could not give away what he did not have: ordinary and criminal jurisdiction over the Indies. In February 1558, Mercado de Peñaloza, *oidor* (judge) of the Audiencia de Lima, suggested granting perpetuity without jurisdiction to the encomenderos of Peru. He felt that the smaller encomenderos would be unable to afford the higher price for jurisdiction and feared that granting jurisdiction to the encomenderos would place the Indians at their mercy with hardly any oversight from the crown.

The encomenderos contended that assignment of the encomiendas in perpetuity, with added jurisdiction over the Indians, would be beneficial for all stakeholders in the colonial enterprise. Greater certainty surrounding their sons' patrimony would *allow* the encomenderos to make capital investments in bettering their charges' living conditions and those of the lands that the Indians worked. These greater investments in work and capital would lead to more prosperity and thus greater tribute in the king's coffers from growth in commerce, agriculture, and mining. An added benefit would be the guaranteed loyalty of the encomenderos. Their loyalty would translate into real savings for the king's treasury since they would happily provide labor and funds for "pacifying" any future rebellions; there would be no need for the king to pay for a standing army in the Indies because the loyalty of the encomenderos would be secured in perpetuity. The irony of these assertions was not lost on either party in the negotiations, as the memory of Gonzalo Pizarro's rebellion in 1544 and its material ravages was still fresh.[10] Moreover, the vecinos residing in Cuzco and Potosí, the richest agricultural and mining centers of Peru, were happy to refresh the king's memory with periodic uprisings in support of perpetuity.

According to the proponents, the renewed prosperity of the encomiendas granted in perpetuity would also provide an incentive for encomenderos to pay the tithes from the Indians' tribute for their evangelization on time, implicitly conceding the truth to charges of laxity in payments that had infuriated many members of the clergy. Their argument thus implied that the lack of perpetuity had created uncertainty that *had not permitted* the encomenderos to properly look after their Indians held in encomienda. Interest, which, as argued in chapter 1, had been vaunted as the price of peril, as a necessary evil, had its own raison d'être. Interest personified made its own claims on the economy of the encomienda in order to secure capital investments. Uncertainty surrounding lease terms for Indigenous labor generated, and legitimated, the unrest of the encomenderos.[11] Yet much of the unrest was caused, ironically, by the insurrections of encomenderos demanding the perpetuity of the encomienda.

When the curacas learned of the commissioners' impending visit to Peru, they decided to meet in Lima in July and August 1559. There they named their advocates, Las Casas and Santo Tomás, both of whom were in Spain at the time.[12] The curacas drew up the following counteroffer to the proponents of perpetuity.

1. After the living encomenderos died, their benefits would return to the crown and no encomiendas would be distributed thereafter.
2. No encomendero or member of his household could ever enter an Indian settlement under any circumstances.
3. Indios in *corregimiento*, that is, indios who paid labor tribute to the crown directly, would see their tribute reduced by half.
4. Each Indian would pay tribute according to his ability to work or pay.
5. Settlements with reduced populations would be incorporated into larger towns so as not to be burdened with high taxes that were based on an earlier census of a larger population.
6. Issues of general interest to all Indians were to be discussed by a general assembly of representatives as in the times of the Incas.

7. The curacas, or *señores principales*, would not be obligated to work and would receive a coat of arms or an empresa.
8. Indian lands would no longer be given to the Spaniards.

The terms sought to create an Indigenous aristocracy (curacas, or señores principales) distinct from the Indians both in livelihood and in self-fashioning. Indians work; curacas do not work. Paradoxically, those who do the negotiating would define themselves as *ociosos* (idle) and receive the empresa (enterprise/sigil) as a sign of their class privilege. At the same time, the curacas advocated for a reduction in tribute, making it commensurate to the Indians' ability to pay. They made this demand on behalf of the Indians directly under their jurisdiction and those rendering tribute to the Spanish crown in corregimientos. The fifth proposition, the consolidation of Indigenous settlements, seems to prefigure Viceroy Toledo's policy of reducciones that centralized Indian populations beginning in 1570. However, the demand to account for the losses in population, and thereby reduce the amount of tribute, speaks to the curacas' traditional roles of protectors of the ayni, inter-ayllu reciprocity, and the lives of their kin.

It is the sixth proposition—the convocation of a general assembly—that deserves our undivided attention. What could this possibly mean? How were these assemblies organized during the time of the Incas? Also, could a new political structure—as proposed by the curacas—enact a remedy for the injuries of the Spanish Conquest? Though the *Apologética historia sumaria* had not been published, members of the Lascasian network shared the conviction of Las Casas, made manifest in the *Apologética* but also in *De Thesauris* and the *Doce dudas*, that the discovery of the Indies by Spain could not be qualified as providential. However, the demand for reparations seems to be at odds with the curacas' bid for (limited) sovereignty in accordance with the terms of venture capital.

The Petition of 1560

In representation of the "caciques and señores naturales y sus pueblos de las provincias de aquel reino que comúnmente se llama el Peru" (curacas and natural lords of the provinces of that kingdom commonly known as Peru), the curacas, and Friars Bartolomé de Las Casas and Domingo de Santo Tomás in their stead, made the compelling offer to Philip II. From the first line of the text, the curacas challenge the Spanish "in the name of the Indians" by promising to outbid the Spanish no matter what their counteroffer might be. Vowing to give 100,000 ducats more than anything else the encomenderos might have offered or would offer in the future, the curacas would instigate a bidding war. The text eschews any pretensions to a shared love or interest with the encomenderos. Instead, it accuses the Spanish—always *los españoles*, never the encomenderos or vecinos—of negotiating in bad faith, and it declares, without any caveats, that there can be no shared interest or love between the two parties. The defining characteristic of their relationship is one of debt and liability, not love or friendship. The references to nationality, and omissions of the encomenderos' title or occupation, reinforces the argument that runs throughout the text: the other bidders do not belong in Peru. They are foreigners who have torn at the social fabric of native societies with their greed, they have rebelled against *their* king, and they are not to be trusted as they do not bid, or do the king's bidding, in good faith. Yet the constant allusion to their competitors—with whom the curacas claim to have no shared interests—as Spanish begs the question: isn't the king Spanish? The short answer, according to the curacas, is no.

In their petition, under the guidance of Santo Tomás and Las Casas, the curacas allude to the long-standing Roman distinction between citizen (*cives*) and foreigner (*peregrinus*) in order to impute a foreign character to the Spanish in Peru. The curacas, whose autochthony is a point of pride, argue in favor of an empire of Christian culture much in the same way Cicero had asserted the existence of Roman citizenship and urbanity beyond Rome in *Pro Archia Poeta* (62 BC), which had anticipated universal Roman citizenship for all free residents within the confines of the empire. Citizenship along the lines of Cicero's arguments would be promulgated by Caracalla in AD 212. The Ciceronian interpretation of Roman citizenship contradicted the earlier practice of jus gentium, which had been applied to *peregrini* (foreigners) during the Republic. Jus gentium, as applied to "foreigners" (those who were not Roman cives), had made contradictory proposals. First, all peoples had a right to govern themselves by their own laws and customs. Second, these laws had to follow universal and natural laws. Third, foreigners (in Rome) and natives (in conquered territories) received equal treatment under jus gentium, which was different from the laws governing citizens.

The curacas' petition to Philip II draws on the distinction between cives and peregrini in order to transcend it. According to the curacas, and their Dominican advocates, Spanish subjects have been so far away from their king, both in distance and in time, that they have become lawless; moreover, the full incorporation of the curacas and their peoples into the crown of Castile would resolve the precarious status of the Spanish vecinos without encomienda. As seen above, the number of encomenderos vis-à-vis the nonencomendero Spanish population was but a small percentage of all Spanish subjects living in Peru. The Spanish vecinos without encomienda had not lost hope that they would one day receive the desired labor tribute in the form of the encomienda. If the crown incorporated the curacas, however, the descendants of the encomenderos would find themselves like many Spanish men who did not take up a trade but instead wandered, offering their skills at arms for sale: vecinos who lived as vagabonds, encomenderos who thought of themselves as kings. The uncertainty surrounding the perpetuity of the encomienda was untenable, especially when Spanish subjects were acting like kings in a strange land.

> [Con la incorporación] cesarán los bulliciosos y malos motivos y orgullosas soberbias y ambiciones que los españoles, teniendo indios, cada hora tienen y les nacen para rebeliones, porque cada uno estima de sí poder ser rey, por la libertad grande que *allá* han conseguido, por estar tan lejos de *su* rey. Y para asegurar este peligro, va la vida que allá no haya español poderoso; y esto saben bien los que cognoscen aquellas tierras y la presunción que en ella cobran los españoles. (Las Casas and Santo Tomás 1957, 468; emphasis mine)

> [[With incorporation] the rowdy and bad motives of the prideful arrogance and ambitions of the Spanish, who have Indians, which they show at all hours in their penchant for rebellions—because every one of them thinks they could be king, because of the great

liberty that they have procured *there*, because they are so far from *their* king—will cease. And to ensure against this danger, there is no way that there can be powerful Spanish over there; and this is well known to those who know those lands well [and see] the pretensions of the Spanish.]

The petition was presented in person by Las Casas and Santo Tomás in Valladolid, though the text itself had been redacted in Peru. However, the deictic markers (*allá*, *su* rey) point to Peru from the *origo*, or originating point, of Spain, and thus underscore the marginal status of "los españoles" who live without the law of *their* king "because of the great liberty that they have procured there" and without the law of the church. To act in bad faith (*malos motivos*) and full of prideful arrogance (*orgullosas soberbias*) was a mortal sin;[13] to do so in a land where preachers were denying encomenderos absolution was reckless. To speak for the curacas from Spain about the ambitions of the Spanish in Peru was to accuse the Spanish of impertinence, in literal and figural terms: they do not belong *there*; they belong *here*, with their king, because they have demonstrated that they are unable to behave properly in remote lands, far from their king in Spain. Granting perpetuity to the encomenderos would make them kings in a foreign land, and Philip II would be king in name only, and just barely, of the roads (*no queda más rey ni señor que de los caminos y aún esto le quitarán*) (468).

The curacas and their advocates recur to the epideictic tradition ("pointing," figuratively) of the orator in an *apolitical* setting: encomia, invectives, and literary portraiture all belong to the epideictic tradition as elaborated by Aristotle (*Rhetoric* 1.9), Cicero (*On Oration* 11.84), and Quintilian (*Orator's Education* 3.7). Properly speaking, these classical rhetoricians proposed rhetorical strategies for oratory that were not typical of legislative or judicial processes. As such, the epideictic is something of a mixed bag but by definition refers to a rhetorical trope and a gesture used by a speaker intent on reifying, even further, the object of his speech. Thus Quintilian emphasized that signaling a figure in order to emphasize its strong or weak points could be employed equally for men or inanimate objects. However, in the classical epideictic tradition, the orator "pointed out" the moral and ethical qualities or inferiorities that a man could possess; his distance from the speaker was not, in itself, a sign of vice or virtue. Moreover, indicating could only be done by an orator who was invested with authority by his public. Las Casas and Santo Tomás, however, invest the trope of *deixis* with a political argument for belonging. The Dominican friars' deictic self-references emphasize their indigeneity to Spain; the curacas, in representing the Indians, speak with the authority of autochthony *from* Peru. The Spanish in Peru are neither here nor there, with no sense of belonging and thus no rights to pertinence and tribute from the land and its peoples.

By alluding to "los españoles" as if *they* were the peregrini in the Spanish Christian empire, the curacas made a bid for Christian centrality in the Viceroyalty of Peru. Their contention that a Christian empire must have a leader who transcended local identity not only functioned within the universal aspirations of the *Requerimiento* but also acknowledged the realpolitik of Spain under the rule of German princes, the Hapsburgs, and their claims to the title Holy Roman Emperor. Charles V—Charles I of Spain—had abdicated his reign as Holy Roman Emperor only two years earlier and had passed on the rule of Spain and its colonies to his son Philip in 1556. Thus bankruptcy, succession, and the fragmentation of the territories held under Charles I of Spain came to a head within two years of the imperial abdication. The curacas' insistence on treating Philip II as a Christian king and referring to their rivals as Spanish sustained the split in identity between Christians and Spanish that Las Casas had promoted so forcefully in the *Brevísima relación*, published less than a decade earlier. Paradoxically, the would-be partners in the negocio at hand, in the bidding taking place in Spain but conducted from Peru, were treating the new Spanish king as a Christian emperor by alluding to an imperial title that had not been passed on in perpetuity from father to son. The rhetorical (and pecuniary) generosity of the curacas and their Dominican advocates would require a reciprocal demonstration of good faith, but in kind, from their Christian king: recognize us as your fellow Christians and loyal subjects who only possess local aspirations to labor and tribute.

The encomenderos accused the curacas of bluffing and negotiating in bad faith: where would they procure so much gold and silver? The curacas give a hint at the source of their prosperity, which would become the defining topic of Las Casas's *De Thesauris*.

Y porque en aquella tierra hay muchas sepulturas que tienen grandes riquezas, y éstas no las quieren descubrir los caciques porque no les tomen sus riquezas y tesoros los españoles, que mande Su Majestad por edicto público que *ningún español toque en ella en descubriéndolas los indios*, y de todo el oro y plata y piedras preciosas quieren dar a Su Majestad la tercia parte, y que a ellos les quedan las dos (Las Casas and Santo Tomás 1957, 468; emphasis mine)

———————————

[And because in that land there are many burials with many riches, which the curacas do not wish to locate [*descubrir*][for fear that] the Spanish will take their riches and treasures, [they require] that His Majesty order, by public edict, that no Spaniard touch [these burials] if these are discovered [*en descubriéndolas*] by the Indians. [The curacas] wish to give His Majesty a third part [of this treasure] and they would keep the other two parts.]

The curacas make several claims in this short paragraph: secret knowledge of hidden treasure in yet-to-be-discovered burials, ownership of both the knowledge of their locations (intellectual property) and the treasure itself (physical property), and a bid to increase the king's carried interest from the customary fifth (quinta) to a third (*tercia*). The curacas also acknowledge a conflict of interest with the indios, who share knowledge of these locations, though the curacas insist that the Indians' knowledge is not proprietary; in the curacas' view, the knowledge of the indios is illegitimate. Fearing that the indios might share knowledge of these burials with the Spanish, in effect, the curacas seek an edict from the king to preempt the Indians from sharing intellectual property with other would-be usurpers (the Spanish). Does the curacas' recognition of possible conflicts of interest with the indios in the matter of buried treasure undermine their advocacy (and thus their petition) made "in the name of the Indians of Peru"? The petition bears the fault lines of "real political subjecthood" in communities subjected to Christian missionary work, a theme I explored in chapter 3.

The curacas and their Dominican advocates defined themselves as a *mancomunidad*, not as a sociedad, and the distinction may be an important one. Rather than associate for a shared interest in material gains, the members of the mancomunidad exact reciprocal obligations of one another around a specific goal. Often, "mancomunidad" is the term preferred by translators into Spanish of the English "commonwealth." According to the *Diccionario de Autoridades* (1724), mancomunidad is "la unión que dos o más personas se obligan al cumplimiento o execución de una cosa. Latín. *Communitas. Communis obligatio, vel in solidum*" (the union of two or more people who commit themselves to the performance or execution of something. Latin. *Community*. The common bond, or the whole [*in solidum*]). The *solidum*, in this definition of mancomunidad, insinuates a synecdochic relationship between obligation and community when it is understood as "whole." However, solidum can also mean gold coin, money (in its plural form, *solida*), or salary. Are the terms of communal obligations set by bonds in capital? Are communal obligations synonymous with money? Is community synonymous with debt? The exact trope is difficult to define in the rendering of mancomunidad by analogy to currency in the Latin examples offered by the *Diccionario*. Yet by invoking "mancomunidad" in their petition to Philip II, and in their willingness to receive less tribute from the Indians, the curacas suggest a desire to define their community without the strictures of capital even as they bid within its terms.

As opposed to the societas, which is structured around a partnership of shared interests and intersubjectivity (recall the etymology, *inter esse*), the mancomunidad of the curacas defines itself as a community of persons committed to one another's welfare. Moreover, a mancomunidad of cities or of towns need not share territorial cohesion but aspires to a community of bodies. Though the petition is often referred to as an effort by the curacas of the Mantaro Valley in central Peru, because most signatories were of Huanca ethnicity, curacas from areas as disparate as Cuzco and Chachapoyas also participated in the effort. Thus, a mancomunidad of curacas, by definition, displays an interethnic level of organization that speaks with one voice (*prestan voz*) for a sole and finite purpose. But how do you procure consent as an advocate for mancomunidad? Is an exchange implied in the *lending* of one's voice to a common cause? Is it *gratis*, that is, charitable?

If the curacas' recourse to the term "mancomunidad" may be ambiguous, it indicates, nevertheless, the political consciousness behind its invocation, as the curacas outline their sixth condition to Philip II. Their sixth "condition" for payment of at least two million ducats for full incorporation into the Spanish crown was a return to pre-Hispanic practices of self-rule based on principles of consent and dissent.

> Lo sexto, que cuando *no* hobieren de tratar los negocios generales tocantes al estado de sus repúblicas, que se convoquen procuradores de los pueblos y sus comunidades, para que lo entiendan y consientan si fueren cosas útiles, o den razón de lo contrario, como lo solían hacer en tiempo de sus reyes ingas, y se acostumbra en las Cortes *acá* de España. (Las Casas and Santo Tomás 1957, 466; emphasis mine)

> [The sixth, that when they are to discuss the general business of the state of their commonwealths [*repúblicas*], that they summon solicitors of the pueblos [towns but also peoples] and their communities, so that they may understand and consent if they be useful things [i.e., the business at hand], or dissent, as they used to do in the times of the Inca kings, and is the custom in the Cortes *here* in Spain.]

The allusion to the Cortes of España with the deictic *acá* generates an ambiguity in the comparison of political and cultural institutions. In the *here* (acá) can be heard the intervention of Santo Tomás and Las Casas in Valladolid. Yet this *here* refers to the Cortes of Spain, not of Castile. In effect, the comparison refers to a false similitude with a nonexistent entity.

The fiction "Cortes of Spain" as a metaphor for the Andean assembly thus blurs the purpose of the degree of autonomy proposed by the sixth condition. The origin of the parliamentary system, known as the Cortes, goes back to the thirteenth century and the power-sharing systems of Castile, León, and Galicia (*xunta*); the Cortes represented the three estates (*estados*)—nobles, prelates, and commoners—in León and Castile, and had decision powers over taxation and the financing of armies. Though Isabel and Fernando reduced the power of the Cortes of Castile and León, making it a largely redundant body in the state apparatus during the sixteenth century, other Cortes, such as

those of Aragon (comprising Catalonia and Valencia) or Navarre retained a significant level of autonomy. Since there were no Cortes of
Spain as a whole (i.e., *in solidum*), the allusion to "acá" in the phrase
"en las Cortes acá de España" begs the question, Where in Spain? Castile? Aragon? Navarre? The ambiguity allows for a spectrum of autonomy in the curacas' bid for incorporation with the crown. In effect,
it is a political demand made in a bid for sovereignty in terms of capital
(literally, a bidding war) that couches its request in ambiguous, juridical
terminology that suggests a *return* to institutions, as in the past, in the
Andes and perhaps in Spain.[14] The curacas' gesture toward a comparable temporality in Spain in their bid for sovereignty in the Andes offers yet another layer of complexity to the assertions made by Kathleen
Davis (2008) about historiography's stakes in periodization during the
sixteenth century. The metaphorical use of a fictional institution, the
Cortes of Spain, to allude to the Indigenous past (*tiempo de sus reyes
ingas*) in a bidding war where degrees of sovereignty were at stake gives
the reified past a specific value; that is, it is part of the "package" of
conditions requested by the curacas in exchange for their payment in
millions in ducats to the king of Spain.

Yet there is also a grammatical ambiguity in the sixth condition. Is
the *no* used as a negative enunciation, or is it employed with an expletory value? If the former, the *no* would limit summons of a pan-Andean
assembly to make decisions as a whole (*entiendan y consientan, o den
razón de lo contrario*) over matters that would not affect the whole.
More likely is the use of *no* with an expletory value in subclauses that
employ the subjunctive mood (as in Latin). This is the only emphatic
use of *no* in the seven conditions stipulated by the curacas as conditions
for their bid. The grammatical structure serves to underscore the importance of this final condition both for the curacas, in continuation of
practices under the Incas, and their Dominican advocates. The "medieval" form of the Cortes, as an assembly where at least three estates
were represented, may have been particularly appealing to Las Casas
and his preference, as articulated by Gutiérrez (1993), for the salt of
the earth.

The curacas contrast their mancomunidad, and their aspiration to
hold regular Cortes, with the sociedad of the encomenderos, a social
organization motivitated purely by self-interest. As alleged by the

curacas, the encomenderos had joined forces out of a shared interest (*su proprio particular interese*). The offer of the Spanish to buy the perpetuity of the encomiendas was a bid to purchase the freedom of the curacas and of the indios and make slaves of them (*en cautiverio perpetuo*). Yet their freedom was unalienable, the curacas argued; it could not be bought and sold. The encomenderos wished something unnatural, "to enslave peoples that are free" (*de pueblos y gentes libres que son, hacelles esclavos*) (466).

In their petition, the curacas make their bid from a discursive space outside of the metalepsis of love interest. There is no love lost or shared interests that bind them to the Spanish, they argue; to the contrary, the Spanish are their enemies, for these foreigners would make slaves of them, buying their freedom for their own usufruct and toward the fulfillment of their wayward political ambitions. Though the encomenderos had continued to make their bids within the metalepsis of "love interest"—that is, perpetuity would be a win-win-win for all parties involved—the curacas declared that the encomenderos' proposal made on the Indians' behalf was nonsense, by definition, "porque los españoles son siempre del bien de los indios contrarios, por su propio interese" (because the Spanish are always against what is good for the Indians, because [they follow] their own interest) (466). The reasoning of the curacas, when doing the bidding of empire, recalls the distinction Christian/infidel in the conquests of Bishop Diego Gelmírez of Santiago de Compostela in North Africa, explored in chapter 1. The infidels' losses are the Christians' gains. In their petition to Philip II, the curacas have no qualms about acknowledging that the Spanish sovereign's acceptance of their terms would result in total losses for the encomenderos, equivalent to the net gains of the native lords of Peru.

Nonetheless, the curacas sustain the metalepsis of love interest when presenting their case for incorporation in terms of maximized benefits for the Spanish crown. Their counteroffer thus remains, in part, within the discourse of love interest when these profits pertain to the Spanish sovereign. Let us recall that the encomenderos had presented a win-win-win scenario wherein the encomenderos would treat the indios with greater dignity and love because they would be able to do business in a climate of certainty, knowing that they were working toward the future of their heirs. According to the encomenderos, greater certainty for their business would translate into better living conditions for the Indigenous peoples in their care, which, in turn, would mean greater rewards for the Spanish monarch, who would receive more tribute from a more productive workforce. By contending that the indios would flourish better under their rule, instead, the curacas aligned themselves with the Spanish king as fellow Christians who shared the same interests; they, and not the encomenderos, were the better custodians of the Indians' biopower.

Allying with some while glossing over others, the curacas' pretensions heighten the conflicts between the partners of the imperial enterprise. Speaking of the Spanish vecinos without encomienda, they noted that granting perpetuity of the encomiendas to the current encomenderos would give this already problematic population—the majority of Spanish subjects in Peru—even more cause for despair, as they would lose all hope of winning a fortune with an encomienda. Yet the curacas offered no answers to the obvious riposte to their argument: How would incorporation of the curacas with the Spanish crown benefit the interests of this large group of Spanish without encomiendas? Their glaring omission of the interest of a competing group increases the friction between caritas and cupiditas that had always existed between the juxtaposed, and later synthesized, terms; what had become the metaleptic habitus of venture capital, the synonymous use of love and interest, was beginning to split at the seams. The demands for a return to an earlier, pre-Hispanic form of the political, the proposal of an alternative form of self-government, indicate a weakness in the persuasive force of venture capital. It reminded the sovereign that the habitus of venture capital was a novelty in contrast to past practices. Though the terms of the petition were spelled out in the terms of capital, the authors reminded their *Christian* sovereign, *it had not always been thus*.

The rhetoric employed by the curacas is also a small mercy for the fama of Philip II. Framed as a petition but structured as a bid for sovereignty in exchange for demands, the text, in other words, could have offered a king's ransom. After all, Philip II had a great need for gold, to recall the words employed by Hernán Cortés in his *Segunda carta de relación*, when he addressed Charles V, almost a generation earlier. The fear of mala fama, of once again losing his credit, had forced the newly anointed king, Philip II, to make the first offer in the late 1550s, thus putting the sovereignty of Peru on the proverbial negotiating table. The synonymous use of *petition* for *bid* and, at certain points in the text, for *remedies* and *reparations* treads ever so softly around Philip II's needs and the curacas' demands. The call for reparations, ever incommensurable to the losses suffered by the Indians, had become another bargaining chip in negotiations with the Christian sovereign.

Is bidding for sovereignty just another way of doing the bidding of empire, of negotiating *within* domination? As argued by Baber (2010), the caciques of Tlaxcala were ultimately successful when they negotiated with the Spanish crown for city status, which held the promise of greater independence in the tradition of the fueros. But how is this success "qualified"? Should we speak, instead, of the "tragedy of success"?[15] Though the curacas spoke in the code of shared interests with the Spanish king, they do not gloss over the language of reparations for losses and injuries. The curacas make one cursory concession to the logic of love interest in their offer to Philip II.

Much like the taquiy ongoy movement that would take hold in southern Peru during the same decade as the perpetuity negotiations with Philip II, the curacas had proposed a political vision for the Andes with a minimal presence of their invaders. However, unlike the taquiongos, the curacas made a distinction between friars, prelates, and encomenderos; their negotiation with Philip II could hardly be called an insurgency. The figure of Francisco Tenamaztle, exiled insurgent from the area known as Nueva Galicia, points to the discontinuity between insurgency and negotiation. As Rabasa (2010, 178–94) contends, the once-naked Tenamaztle, the nomad in the desert, now an exile in Valladolid, must clothe his case for reparations in the discourse of jus gentium and natural law as a Christian, with Las Casas as his advocate; his insurgency, much like his former nudity, does not belong in the annals of history. Las Casas argued Tenamaztle's case before Philip II in the same year that the crown's hacienda declared bankruptcy, 1556. Four years later, bidding for a measure of sovereignty would be grafted onto the language of reparations in a petition sent from the curacas of Peru.

The curacas show more complicity with the terms of negocios, much like Ruíz Medrano's (2010) Mexica or Baber's (2010) Tlaxcalans. Yet the terms employed by the curacas to settle the matter of the encomiendas once and for all are rather unsettling. We hear the voice of Las Casas in the curacas' calls for reparations, yet cannot reconcile these calls with the despair of Las Casas in the *Doce dudas*: the desolation is irreparable; no remedy is commensurable to the loss. At the same time, the curacas' demand for reparations contradicts the gesture of negotiating the terms—in ducats of gold and silver—of the limited sovereignty of Indigenous states under Christian empire.

The curacas picked and chose the elements of empire that they were prepared to accept: the empresa (coat of arms) and the salary of landed nobles in the Castilian tradition, increasing the king's share of buried treasure and mined ore and reducing Indian populations into dense settlements. What had been and remained unacceptable for the curacas was the cohabitation of Indians and Spanish. Could it be that the curacas were invested in a parallel system of government that would later become the *república de indios* and the *república de españoles*? If so, their bid for a república de indios with themselves as the sole mediators between native labor and the king was ultimately rejected.

How can we approach this failed bid with the respect it deserves, listening to the bid in the integrity of its moment, without recourse to the narrative arc of historic development? Were the curacas doing the bidding of empire (obeying it) or just making a bid for indirect rule within the imperial framework? Theirs was clearly not a peasant insurgency, in the paradigm explored by Guha, but their attitude to the encomenderos, at least in this moment in time, veers far from their characterization, by Stern (1993) and others, as the middlemen between the república de indios and the república de españoles. In the decade of the 1560s, prominent native elites were attempting to bid out of this parallel system of government in the colonies. Can we speak of negotiation as a "counterbalance" to domination, as suggested by Owensby (2010)? The more urgent question remains, can negotiation exist *without* domination, when its activity—busy-ness, neg-otium—belongs to the time and activity of capital? And yet there remains the nagging feeling that this question responds to the a priori of capitalism's categories. What remains beyond dispute, however, is that the curacas had waged a bidding war against the encomenderos in terms of absolute hostility. Is insurgency the only way out of empire? Can we speak of a "bidding insurgency" as opposed to a "bidding war"?

If Philip II had accepted either offer, perpetuity or incorporation, it would not have taken effect until after the expiry of the two-lives limit on the encomiendas, established in the (in)famous law passed by Charles V, "Derecho de suceder por dos vidas," in 1536. It is worth noting that the encomenderos had chosen to bypass those who indeed possessed local jurisdiction over the Indians by virtue of jus gentium— the Indigenous elites, or curacas—even after the Consejo de Indias and the king eventually admitted that jurisdiction over the Indians was not theirs to give. Why did they not negotiate directly with the curacas? Both the encomenderos and, later, the curacas engaged in bilateral offers and counteroffers with the crown but did not negotiate among themselves in Peru. If the encomenderos had wanted civil and criminal jurisdiction over the Indians, why didn't they make two offers, one to purchase such jurisdiction from the curacas and another to purchase perpetuity of the encomienda from the crown? It may be rash to venture an opinion based in part on a counterfactual, but the lack of overtures between the encomenderos and the curacas implies, despite the rhetoric of a win-win-win proposal by the encomenderos, that their differences were irreconcilable; that they could not negotiate with each other, that is, do negocios, let alone engage in dialogue; that jus gentium was unalienable; and that the ties between peoples and land could not be bought and sold. So, within one territory, two groups of elites ignored each other and negotiated with the sovereign on another continent.

In other wor(l)ds, the curacas' rejection of the encomenderos as a party to negotiations also rejects the advances of the conquistador as the middleman, or matchmaker, of a loving empire. Unlike the Incas or the Mexica whose creation myths follow the emergence of the first ancestors from distant caves to the foundational moment of an urban center, of the polity that would have authority over different peoples, the curacas had always emphasized their ties to the local huacas. As direct descendants of the local, foundational ancestors, they are the truly autochthonous, since their identity and power are tied to the local *chthonic* power.[16] In initial contacts with their Spanish invaders, some peoples in Mexico and Peru cast their lots in favor of greater local autonomy. Like the Tlaxcalans who had allied themselves with Cortés, and later demanded recognition and greater autonomy from the crown for lending their support to the European invaders, the Huancas in the

Mantaro Valley and other Andean peoples under the yoke of Inca rule in the territory comprising the Tahuantinsuyu similarly allied themselves with the Spanish to defeat the Incas.

Were these alliances motivated purely by strategy or self-interest? As Gose (2008, 36–80) has argued, much like the identification of Cortés with Quetzalcoatl or the native Hawaiians' identification of Captain Cook with Lono in the eighteenth century, the Indigenous of Peru positioned the European invaders within their own origin stories.[17] This native positioning of the outsider as insider not only served to make sense of the upheaval in the Andean experience of the world following the death of Guayna Capac and the events that followed but also contested emergent Eurocentrism by articulating a "politics of connection," which according to Gose, engaged in a "deliberate, counteracting response to racism," a trope that is chraracteristic of Indigenous accounts of colonialism worldwide (20).[18] The Andean trope of the conquistadors as *viracochas* (uncanny ancestors) would have served as an "interethnic collaboration" that led to a system of indirect rule. In this scenario, curaca-led insurrection against the Inca state in alliance with the Spanish invaders would eventually generate the system of Spanish colonialism, but initially, until the crisis of the 1560s, it resembled greater provincial sovereignty.

Why did the curacas abandon their erstwhile partners in insurrection against the Incas? No triangle of negotiation has come to light; negotiations were done on parallel tracks, despite the rhetoric of win-win-win used by the Spanish proponents of the perpetuity of the encomienda. Yet the discourse accompanying these parallel tracks of negotiations, between king and encomenderos, between king and curacas, followed the rhetoric of love interest, the synonymous use of caritas and cupiditas. This was especially true in the texts composed by the encomenderos. Perhaps these parallel negotiations serve to prefigure the Spanish juridical arrangement known as the república de españoles and the república de indios.[19] However, pinpointing an origin for a juridical structure, in this case, the parallel republics, does not mean that our analysis of the same ought to replicate its discursive parameters. By the same token, the widespread use of "negotiate" as a catchall term for actions and practices of Indigenous and other marginalized peoples ignores the actual practice of negotiating, the busy-ness of it.[20] Negotiating with identity is most certainly relevant to the experience of conquest; recall the negocio of Indigenous identity in Isabel of Castile's instructions to Ovando, discussed in chapter 4. To negotiate, first and foremost, is to engage in a practice that negates an unproductive use of time.

The "invaders as ancestors" trope insists on an Indigenous framework for appropriation of the imperial modus operandi and agency in the generation of colonialism in the Andes. It is also an argument that depends, to a certain extent, on the longue durée of politics, belief, and religion in the Andes. By the time the curacas were petitioning Philip II for incorporation, however, this identification of invaders as ancestors seems to have been expendable to the extent that the curacas were requesting a renewed system of indirect rule without the involvement of the conquistadors turned encomenderos. While it is true that the petition was mediated by Domingo de Santo Tomás and Bartolomé de Las Casas, it is difficult to imagine the latter suppressing such an important part of the native lords' belief if the curacas had indeed taken it into consideration.

Yet the curacas' alliances with Christian leaders did not immunize them from existential threats to their sources of power. The regulation of Indigenous marriage according to Christian precepts increasingly undermined a crucial aspect of self-rule until, by the late sixteenth century, the festive consumption of corn beer (*qura*) and the women, married or not, who produced it became associated with sexual excesses, transgressions of Christian precepts on which, paradoxically, the "native" system of labor, and thus the colony, depended (Gose 2008, 140). The curacas' bid for incorporation is no less interesting because they continued to enjoy a pivotal role in the exchange of labor for spiritual tutelage, despite their failed attempt at self-rule in the bidding wars of the 1560s. At the same time, their adherence to Christianity through indirect rule acquiesced to an important limitation on the curacas' power to control the minka through women's labor and marriage. The minka, as opposed to the state-enforced mita, functioned on reciprocal exchanges among ayllus, or kinship groups with a common ancestor; the curacas accessed an excess of labor in these horizontal exchanges by offering an excedent of food and drink, products of women's labor. The issue of polygyny, thus, accounted for the power of the curacas to control access to labor by the state, whether Inca or Spanish. Yet polygyny was illegitimate under Christian doctrine, especially after the Council of Trent reinforced the treatment of marriage (between one man and one woman) as a sacrament. How, then, to resolve the contradiction between Christian tutelage and its material dependency on pagan practice? Moreover, when practiced among baptized Indians and curacas, polygyny was a sign of apostasy or heresy.

The economic and moral values of the system contradicted one another, but Christian hegemony *depended* on a contrary system of moral and economic values. How could the fruits of Indigenous labor in a domestic and political economy that was dominated by polygyny pay for the salaries of the Christian doctrineros? Could the doctrineros simply turn a blind eye to the material contingency of their evangelizing enterprise? Or could Christianity embrace the Indigenous practices that seemed to turn their backs on Christian doctrine? Part of the answer to these questions was provided by the heresy trial of the Dominican friar Francisco de la Cruz, with repercussions for the Lascasian movement after the transitional decade of the 1560s. Fray Francisco de la Cruz's testimony bore witness to the complicity among members of all sectors of colonial society, including the curacas, with these structural contradictions (Abril Castelló 1992, 195). But de la Cruz's delirium also responds to the structure of venture capital's love interest and its contingency on the Andean habitus.

OTHER PROPOSALS: THE HERESY OF FRAY FRANCISCO DE LA CRUZ

Fray Francisco de la Cruz's trial by the Inquisition in Lima, in which José de Acosta participated, gave voice to the fears about what an assembly or Cortes of various Indigenous estates might have enacted if their bid with Philip II had succeeded.[21] Francisco de la Cruz was a Dominican friar who arrived on the same ship that brought fray Domingo de Santo Tomás to Peru at a time when, as contended by Bataillon (1966, 323), Peru suffered from rapid "criollization." Yet the "criollization" process began almost as soon as Hernando Pizarro returned to Peru from his trip to Spain to deliver the massive quinta to Charles I. Armed with a *cédula* (edict) from Charles V, which promised encomiendas granted in perpetuity, and a series of cédulas, from 1534 to 1536, that gave Francisco Pizarro the power to name his successor to the governorship of Peru, the encomenderos cannot be faulted for believing that they *belonged* in Peru (Lohmann Villena 1977, 49–50).[22] The promise of legal ties to Indigenous labor in perpetuity, combined with marriages and *mancebazgo* (concubinage) between conquistadors and Inca noblewomen, had contributed to a burgeoning sense of autochthony among the conquistadors.

The prophecies and ambitions of de la Cruz offer a monstrous refraction of the Spanish Conquest, its insurrections, and its doubts. The *Requerimiento* had asserted that the pontiff could move his seat of power anywhere in the world; de la Cruz prophesied that Lima would be the new seat of Catholicism, and his son with doña Leonor de Valenzuela, Gabrielico, would reign as new pontiff with de la Cruz ruling at his side as temporal sovereign. The union of temporal and spiritual powers represented in the chivalric fiction of Fernández de Oviedo's *Claribalte*, discussed at length in chapter 3, errs and is recentered in this other New World fiction, also the center for the end times.

Just as Acosta would apply a cost-benefit analysis to martyrdom in the Indies twenty years after the *escrutinio* (investigation) of Francisco de la Cruz, the Dominican heretic would apply a similar reasoning to massacre during his trial for heresy. For de la Cruz, it would be better "que no vayan a la conquista pocos españoles, porque los indios, viendo a pocos españoles, se atreven a acometerles y a resistirles más, y es ocasión que se haga mucha carnicería en los indios" (that more than a few Spaniards go on conquests, because when the Indians see few Spaniards, they are reckless in their attacks and resist them even more, which gives way to the butchery [*carnicería*] of the Indians) (Abril Castelló and Stoffels 1996, 621). De la Cruz perceived that a larger group of armed men would have dissuasive power over the Indians. Thus coerced consent would be worth the extra expense *and* save more lives. Though de la Cruz aims to convince his audience that the end of days is approaching, he does not stray far from the economics of conquest. For suggesting that the church had its own Casa de Contratación, Francisco de la Cruz turns on himself, denying his voice as he utters his condemnation, charging God with the task of damning God's church.

Y así ahora, aunque conoce que está obligado a decir las cosas que
ha dicho del Papa y de la Iglesia, y tuviera por pecado mortal de-
jarlas de decir, pues que se lo ha dicho Dios para que se enmienden,
por otra parte las dice con vergüenza, hablando como hijo que es
de la Iglesia Romana, y viendo que habla contra sus padres y cabe-
zas. Y por esto dijo *no quiero decir que lo digo yo.* Y que siente que
es así por lo que Dios le ha declarado: Que la Iglesia Romana,
adonde está el sumo poder espiritual, *trata las cosas del gobierno
espiritual como por vía de contratación.* (Abril Castelló and Stoffels
1996, 829–30; emphasis mine)

[And so it is now, though he recognizes that he is obligated to say
[repeat] the things that he had said against the Pontiff and the
Church, for silencing them would be a mortal sin, for God has told
him these things so that [the Pontiff and Church] be reformed, he
says these things in shame, speaking as a son of the Roman Church,
speaking against his fathers and leaders. And for this reason he
said, "I do not wish to say that I say this." And he feels this way
because God has declared to him: the Roman Church, where the
great spiritual power resides, *treats the matters of spiritual government like a negotiation for contracts.*]

The speech of reform is motivated by shame, by a self-proclaimed son
speaking against his fathers. Ashamed of what God impels him to declare, he fears the mortal sin of remaining silent. The church, he confesses, *trata* (treats but also trades) the things of spiritual government
as if by contract, as if it were another Casa de Contratación. He dares
to speak like a Lutheran, he continues, to preempt Lutheran speech
against the church. His heresy would preempt heresy. Following his
death at the stake, the Inquisition retried his case to verify if he was indeed "hereje" or "loco." This binary proposition raises questions remi-
niscent of the conundrum posed by Aquinas of the child in the cave
discussed in chapter 4: Is rational rejection of the Catholic faith even
possible? Is radical reform of the faith a proposition for the mad?

   The prophecies of Francisco de la Cruz manifest the imperial impulse toward synthesis, a vast monster that would consume all sources
without prejudice in an effort to reconcile all contradictions by any

means. The strain of the enterprise creates a discourse full of fissures. In the pretensions of de la Cruz for his native-born son, Gabrielico, can be heard the echoes of the pretensions of Gonzalo Pizarro and the chauvinism of Fernández de Oviedo. Francisco de la Cruz, heretic, accuses the Indigenous of apostasy; they are the descendants of the lost tribes of Israel who turned against God and chose to worship the devil. As apostates, they may be *conquered*, an argument that hews to the logic of Vitoria's legal titles of war while straying from the position, more generally accepted, that the Indigenous had no prior explicit knowledge of the Mosaic covenant or the law of grace before the arrival of the Spanish. However, his advocacy for polygyny and adultery not only made him a heretic and a sinner, but also an apostate, though one subsumed by the practices he enjoyed and observed among his parishes in southern Peru and Lima. This turning away from his native Spain and from orthodoxy resounds in the preponderance of apostrophe in his testimony on the prophecies of the angel that spoke to María Pizarro.

Turning away from God, betrayal, adultery: these are all tropes for apostasy in the Old Testament (Jeremiah 3:1–3; Ezekiel 6:9). The forced reconciliations of love interest became distorted, doubly confounded; de la Cruz's insistence on aligning trope and prophecy create an "uncanny" coherence in his message.[23] It is almost as if one can recognize his heresy solely on the tropes of apostasy that he impugns against the native peoples of Peru. From de la Cruz's delirium—the deviation from the furrows of cultivated soil, what is also known as the area without the nomos—he accuses others of stepping out of bounds, and yet he is incapable of imagining the expulsion of Christianity from Peru. Even so, de la Cruz is the first to admit that he knows not whether the angel that speaks to him through María Pizarro is good or evil. It matters not; he must obey.

Beyond his injunctions to practice polygyny or the twisted reasoning behind his recommendations for armed campaigns against the Indigenous peoples of the Andes, it is this uncertainty about the origin and *value* of the oracular voices that Francisco de la Cruz turned to that so confounded his inquisitors. De la Cruz's tenacity was the object of begrudging admiration from his inquisitors, including Acosta. His belief that *these things had to be done* belonged in an amoral system whose imperatives he nonetheless felt compelled to obey.

How could de la Cruz have obeyed those voices if he was *uncertain* about the moral provenance of these orders, good or evil? If it had not been so viscerally performed, de la Cruz's delirium almost would have seemed a parody of probabilism's moral theology. Though he questioned the sanctity of his source, de la Cruz nonetheless offered his wisdom to the theologians who would be meeting at the upcoming Third Council of Lima, so that everyone would understand that "estos *negocios* son de Dios" (this *business* belongs to God) (Lohmann Villena 1977, 890). At the same time, de la Cruz's unquestioning obeisance to the voices' imperatives reflected a monumental shift in consciousness to the widespread delirium, in its etymological sense, (en)forced by venture capitalism on a global scale but experienced in the daily transgressions of Spanish and Andean systems of value. If we were to replace "market" for "angel" in the prophecies given by de la Cruz's testimonies, we find something eerily close to modernity's subscription to free market capitalism.

In a Manner of Conclusion

In January 1562, in the town of Mañaques on the outskirts of Lima, the curacas chose their own legal representation to make a counteroffer to Philip II. In addition to Domingo de Santo Tomás, who had just returned to Peru from Spain, where he had recently published his *Vocabulario* and *Gramática* of Quechua, Gerónimo de Loaysa, archbishop of Lima, and fray Francisco de Morales, a Franciscan, were in attendance. Bartolomé de Las Casas was named as a replacement, along with fray Pedro de Cepeda, prior of the Augustinian monastery in Cuzco, among others.

Licenciado Polo de Ondegardo and Friar Santo Tomás traveled together from Lima to La Plata to argue their respective cases for perpetuity of the encomienda and incorporation of the curacas with the Spanish crown, respectively. Both solicitors would later claim in their correspondence with Philip II and the Council of the Indies that the Indians overwhelmingly supported the offers for incorporation, or perpetuity, of their respective clients. Though Las Casas's advocacy of the curacas' offers supports the claim made by these native elites to be the *true* advocates for the Indians of Peru, even their petition concedes, in part, that they could not ensure the consent of *all* Indians to their enterprise, at least not without the coercive power of Philip II in the matter of the huacas and the ongoing conquests for their treasures.

What, then, of the figure of Indian consent? It has been met with its own monstrous likeness in the Andes. To this day, Andean collective memory recalls the horror of conquistadors rendering Indian fat for skin salve. Known as the *naqaq* or *pishtacos*, these vampirelike creatures appear during periods of chaos to kill native peoples for a commodity that fetches a high price in international markets; they make their product of human grease but never sell it locally. The invaders as vampires have found different uses for human fat over the centuries: candle making or smelting bells for churches, greasing railroad locomotives, lubricating jet gears during the space race, or seasoning dishes in Lima's fine dining scene. It is a powerful fiction elicited by the remnant, or the real political subject during the "Manchay Tiempo," the time of fear, as narrated eloquently by Nelson Manrique in *El tiempo del miedo* (2002). Never reduced to a majority or a minority, those who reject the metalepsis of venture capital in taking possession of their lives, yet feel its claims on their life force, conjure up the naqaq.

In his *Fábulas y ritos de los Incas*, written in 1573 to narrate his experience as an extirpator of idolatries during the turbulent 1560s, Cristóbal de Molina dismissed the fears of Indians as fabricated nonsense. For Molina, the Incas of the Vilcabamba insurgency had invented the naqaq in order to generate enmity among the Indians toward the Spanish encomenderos. He argued that there was a real danger to this fiction because it drove fear into the Indians, who had become "muy recatados y se extranaban de los españoles en tanto grado que la leña, yerba y otras cosas no la querian llevar a casa de espanol, por decir no los matasen allí dentro para les sacar el unto" (very shy, and they hid from the Spanish so much that they would not take lumber, grass, and other things to the house of any Spaniard for fear that they would be killed inside to extract their grease) (Molina 1989, 95). That the Indians feared him and all the Spanish in Peru alarmed Molina. The Indians' fear raised the specter of absolute hostility and with it an end to the negocio, the busy-ness, which was the business of Indian labor in exchange for spiritual tutelage. However, Molina's dread also reveals the contingency, as we have seen before, of the evangelizing enterprise on Indian consent. As for the veracity of the claims about the *unto del indio*, one of Molina's contemporaries, Bernal Díaz del Castillo, who was a vecino of the empire residing farther north, may shed some light on the trope's material contingency. In his *Historia verdadera de la conquista de México* (1538/1632), he alludes to the practice of rendering Indians' fat for the purpose of curing the wounds of his associates in conquest (Díaz del Castillo 1984, 2:233).

Over the long term of Andean experiences with violence for corporate profits, there has been no consensus as to the existence of the naqaq, and yet the naqaq as a trope, embodied by the very bodies for whom the horror was felt at the uses and (ab)uses of human flesh and its toil, traded as a commodity on world markets, has become second nature. An unnatural creature, much like capital breeding capital in the Scholastic trope for interest, the naqaq gestates within the empire's reliance on the (ab)uses of biopower and terrorizes by rendering the human body for consumption and production of another embodied trope: empire and its scalability.

Epilogue

(No) Exit: The Maroons of Empire

> Symerons, a black people who about eighty years past fled from
> the cruelty of their Masters the Spaniards, and grew since into a
> Nation under two Kings of their own, one inhabiting Westward,
> and the other East in the way from Nombre de Dios to Panama.
> —R. B., *The english hero*

Over millennia, venture capital has been practiced under the guise of
different names: societas, commenda, triple contact, and merchant
capitalism, to mention a few. My approach to "conquista" has been
contingent, all this while, on an anachronistic term: venture capital.
Privileging this term, because its name references the performative
aspirations of its past and future practitioners, does—admittedly—
approach the caricature of Scholastic nominalism that Rabelais satirizes
in the *Gargantua*. Yet the fact of venture capital's happy translation, all
these years, so many names, so many practitioners on a smaller or
larger scale, points to the successes of a practice that has performed its
figurative and fictional tropes exceedingly well in various cultural contexts. Alluding to White's many works (1985, 1987, 1999) on the tropes
of writing history, Genette (2004, 13) claimed that the pretensions of
metalepsis in historical writing showed more audacity than the same
trope used in fiction; after all, nobody can truly control the past or the
future. The metalepsis of venture capital, in the performative texts and
enterprises of Iberian conquests, laid claim to the future in its gestation
of legal fictions that haunt us to this day.

Yet one salient aspect of venture capital, in its original guise, has been lost in translation. In chapter 1, I argued that two exceptions to usury allowed for the practice of venture capital enterprises on a large scale in the Indies. Sharing risks to life and property underwrote seafaring expeditions in moral terms; the risks taken by partners in an enterprise could justify extraordinary, *unnatural*, usufruct as the stakes were so high. These two exceptions to usury and jus gentium were enough to drive a wedge between use and property values. The conjunction of capitalism and evangelization led material and spiritual wealth to be confounded, in concept and in practice. Pursue your own interest, lovingly. Risk your life to win a fortune: the labor from new Christians. The contradiction between caritas and cupiditas, though never far from view, was submerged, in part, by the immediacy of the exception in these ventures into inhabited terrae incognitae: financial and labor partnerships and the imperatives of free movement for trade and evangelization.

As the biopolitics and bioeconomy of Indigenous labor were scripted into imperial institutions, unabashedly, the audacity of the conquest's metaleptic habitus took on a new guise: the establishment of rules and protocols to limit moral and material risk taking. What had justified, morally, the immoral and unnatural pursuits of usury had become dispensable as a source for legitimation after the fact. As argued in chapters 2 through 4, the increasing importance of love rhetoric emphasized Indigenous consent in the law and contracts of conquista, to the point that conquista itself became an outlawed term in 1573. Yet the premises behind the economy of liberty and productivity, as seen in Isabel of Castile's instructions to Ovando discussed in chapter 4, did not change with the 1573 *Ordenanzas*.

My intention has not been to question the sincerity of the sovereign's struggles with her conscience. The stated goals of paz, amor, and caridad in relations with new Indigenous subjects were all too real. Outlawing the word *conquista*, paradoxically, was a sign of the empire's heavy investment in the metaleptic habitus that made conquista possible in the first place. Conquista was risky business, descubrimiento perhaps less so, though the terms, as Gibson (1977) pointed out, were largely synonymous.

Without risk to life, limb, and property to support risky business (i.e., usury) as a common, moral denominator, the metalepsis of venture capital became self-perpetuating: its habitus was anti-risk. As the metalepsis of love interest became second nature to empire and its agents, risk and unruliness were themselves considered *unnatural* to the desired order. Thus Father Castro and the encomenderos could propose their arguments for perpetuity, couched in concerns for risks to their business, in all sincerity. They had to do their business (with Indigenous labor) with peace of mind; the risk of material losses (from the loss of the encomienda) did not allow them to make the necessary investments in Indigenous well-being. Risk, once integral to the metalepsis of venture capital, was externalized. However, risk continued to justify love interest but from a place outside of the legal tautology that relied—it was declared in 1573—on peace. *In order to show our love to the Indigenous subjects placed under our tutelage*, the encomenderos and their allies in the church argued, *our interest must be ensured, in perpetuity.*

How can one counter such arguments leveled in all sincerity when the categories governing comportment have been confounded on such a grand scale? The moral stakes in telling and retelling the recent past with a view to the immediate future could not have been higher. In response, Las Casas told and retold the pre-Hispanic past and the Iberian conquistas, from Africa to Goa to the Spanish Indies, in order to remember and reiterate the real sacrifices made in the name of legal fictions. What had been cast as an academic debate was recast by Las Casas and his followers as a *dubium*.

This doubt haunted José de Acosta even as he attempted to cast it aside and promote cupiditas as a model for caritas. For Acosta, the relationship between Christian love and empire was one of synecdoche, the part for the whole. Yet power was refracted along the fault lines of investment that divided up the proverbial pie. In his conceit for imperial power, Acosta left little *actual* power for the church as he understood its relation, fully subsumed into the Spanish or Portuguese imperial projects. He also divested real power from Christian love; power, for Acosta, as I showed in chapters 1 and 4, existed in proportion to capital investment. If one did indeed reap what one sowed, as Acosta repeated tirelessly, his proposals for the liberation of the Indies' barbarous peoples undermined his own authority as a member of the church. The contradictions inherent to the synonymous, metaleptic treatment of love interest led Acosta to upend the moral hierarchy of caritas and cupiditas in favor of interest. According to Acosta, biopower expended in favor of cupiditas had borne more fruit than that spent on caritas. The utility of caritas qua caritas became Acosta's main source of doubt.

Determining the nature of the conquest, of its legal status, whether it could be qualified as a Christian enterprise or not, was rediscovered as matters of conscience. The confessional, the preferred means of conversion for the Dominicans, via acceptance and consent of the neophyte, was used to cultivate doubt in all participants of this corporate enterprise. The *Avisos para confesores* in the Viceroyalty of Peru adjudicated complicity to anyone who received material benefits, directly or indirectly, from the injuries made to the Indians. Who would not be implicated after such a far-reaching inquiry into the habitus of the empire's inhabitants? The extent of the devastation, according to Las Casas, was impossible to remedy; the irreparable losses of the conquest, which had damaged the souls of so many of his countrymen, including that of the sovereign, had created an aporia from which Spain had no choice but to retreat. Another prominent Dominican, fray Francisco de la Cruz, would attempt to resolve the contradictions of empire by recentering it with a new political and spiritual see in Lima. This program for reform remained beyond his inquisitors' scrutiny, even after his death at the stake in 1578, to qualify either as madness or as heresy; the Inquisition's repeated attempts to try de la Cruz, even posthumously, perform a kind of neurotic coloniality. Can we infer from this ongoing opposition between heresy and madness that heresy remains a rational option, even when it goes against the law? Where does apostasy begin and heresy end?

Who could make a claim to usufruct without the metalepsis of conquest in the Indies? The doubts raised by fray Francisco de la Cruz and his inquisitors (heresy or madness?) might allow us to imagine other similarly incongruous oppositions: pirate or conquistador? maroon or loyal subject? fact or fiction? Or, rather, to rediscover situations, even "emergency situations," that allow us to recast those binaries for concepts that favor the traditions of the oppressed.[1] The tale of the maroons of Panama and their alliances with the dreaded pirate (and knight) Sir Francis Drake (d. 1596), in the employ of Elizabeth Tudor, is a case in point. Narrated in episodes in Lope de Vega's epic, *La Dragontea*, the descriptions of the maroons, and their actions, underscore the empire's fraught relationship with insurgency, as argued by Guha in *Elementary Aspects of Peasant Insurgency* (1999), to document, rationalize, and suppress. It also exemplifies the tenuous hold of empires on the peoples and the material resources that fueled their hold on power.

At the bottleneck for the transportation of Andean bullion to Spain, runaway slaves (maroons, or *cimarrones*) made alliances with pirates to attack the treasure galleons on the Isthmus of Panama. Francis Drake made an important alliance with these insurgents in and around Nombre de Dios in the late 1560s.[2] This alliance led to his military and economic successes in Nombre de Dios and Cartagena in 1572. Notably, the English apologists of Drake's forays into Spanish "dominions" would emphasize the rights of the former slaves as a people with legitimate rulers. For example, an English source for the episode, known only by his initials R. B., emphasizes the heroics of the cimarrones and also their status as a *nation* in his *English hero* of 1692, quoted in the epigraph to this chapter. More than one hundred years after the maroons' incorporation into the Spanish Empire, the possibilities for an English alliance with this nation remained open, narratively speaking, by omitting Spain's incorporation of the latter-day insurgents. Is the struggle for the recognition of a people's rights another form of consent to the imperial project? Is insurgency the only way out? If so, how can we speak of insurgency without recurring to the figures and fictions of the metalepsis of venture capital and its unrelenting pursuit of scale and scalability?

Lope de Vega's conflicting portrayals of the maroons of Panama point to the limits of form in defining what exists out of bounds. Fomented by greed (*codicia*) in the *Dragontea*, the maroons are motivated by "treachery" and vengeance against their former masters. The epic narrative, however, comes across a stumbling block in its characterization of the insurgents: how to rationalize the subsequent turn of events. Following Drake's and the cimarrones' victories on the Caribbean and Pacific coasts, Spanish colonial officials set out to incorporate the former slaves into Spanish dominion. By 1592, the cimarrones had been "reduced" and relocated to Santiago del Príncipe to live "civilly" (*con poliçía*) and had accepted the authority of the Spanish crown and the church.

They proved their allegiance to the Spanish crown by attacking Drake on his return to the coasts of Panama in 1592, and Drake retaliated against his former allies by burning down their town. By then, the cimarrones were no longer cimarrones (by definition, insurgents) but Spanish subjects whose kings, Diego and Pedro Yalonga, were accountable to Spanish officials in the Audiencia of Panama. Paradoxically, becoming subjects of the crown with its recognition of local rites, customs, and leadership (jus gentium) entailed at the same time an abdication of their identity. Having disparaged the "treachery" of the cimarrones of Panama in earlier cantos, Lope de Vega hailed their valor against Drake by describing them as "almost European" (*como si fueran naturales de Europa*) but also compared their prowess to that of Ottoman soldiers. Once incorporated, the former insurgents are clothed with the mixed metaphors of empire.

How might we qualify these partnerships for profitable violence that are defined per se as outlawed but on which empire depends? Did the maroons show a political consciousness *without* the law? What differentiates a pirate from a conquistador? Or a band of outlaws from a nation? What happens when people, formerly reified as property, become insurgents? And from their insurgency consent to be transformed into subjects? Grammatically but also politically does insurgency exist at the interstices of the figural relation between producer and produced in the metaleptic habitus of venture capital? What are the traditions of insurgency? Can they be transformed into the forms governed by the nomos? Or does the nomos require translation, one that might be, paradoxically, out of bounds?

NOTES

INTRODUCTION

1. Guamán Poma de Ayala 2001, fol. 73. Translations in this study are my own unless noted otherwise.

2. Adorno 1986, 124–25.

3. In this reading, I am not evaluating the juxtaposed images along a false distinction between cognitive and aesthetic interpretations, between, in this case, an allegory that is accurate vis à vis one that "feels right." Rather, my aim is to show how Guamán Poma's allegories present an argument for interpreting what conquest means and how it is signified. See Melville and Readings 1995. In this way, Guamán Poma's visual allegories ask us to consider *why* and *how* we interpret his representations of conquest *aesthetically* or *cognitively* and why we abide by such an opposition in the first place.

4. Repurposing the neologism employed by Latour in *Pandora's Hope* (1999, 257), coined through his recollection that both "fact" and "fetish" share the same etymology, the conquest as *factish* involves both human and nonhuman actors, a double articulation or multiplicity that "is so very real" precisely because it is constructed. "Rhizome-like," as Latour argued, in a nod to Gilles Deleuze and Félix Guattari, "their consequences are unforeseen, the moral order fragile, the social one unstable" (288).

5. Badiou introduces the *event* and the *evental*, his translator's neologism, first in *Being and Event* (2007) and again in the *Logic of Worlds* (2009).

6. Recalling Moreiras's assertion, via Marx and Althusser, that "primitive accumulation is an unthinkable" in the history of the conquest, the question of defining the *origins* of capitalism and the current world-system (Wallerstein) comes to the fore as an impossible possibility within teleological history (Moreiras 2000, 345). And yet if we incorporate Federici's (2004, 11–19) contention that *primitive accumulation*, especially that of the reproductive labor of women, is an ongoing event, then primitive accumulation would no longer be conceptualized as an indeterminacy in its relationship to capitalism, but rather they would

constitute and be constitutive of a double internality to each other. And our own status as "organic thinkers," per Gramsci (1997), of imperial reason in the twenty-first century would once again gain greater urgency (and greater indeterminacy). These questions are explored further in chapter 1.

7. See Maticorena-Estrada 1966, 56–61. The Compañía de Levante (as the southeastern shore of the Isthmus of Panama was also known) is not to be confused with the Levant Company, formed in Britain in 1592 for the purpose of trade with the Ottoman Empire and the Levant.

8. This and Anghiera's other correspondence on the earliest Spanish expeditions were included in the first of his eight *Décadas* and published in 1511 (1.9). All eight *Décadas* were published as *De novis orbo* in 1530.

9. This does not mean, however, that the synonymous use of *caritas* and *cupiditas* went unquestioned. It would not only be rejected by Las Casas, but the two words continued to survive as separate entities in Ercilla's diatribe against greed in the *Araucana* and received an ironic commentary by Cervantes, perhaps, in the love note scribbled on the back of a receipt, by the idealistic, and at times pragmatic, Don Quixote de la Mancha.

10. This is not the place to analyze in detail Frédéric Lordon's polemics with Étienne de La Boétie (1530–1563) and Pierre Bourdieu on "habitus" and "voluntary servitude" with affect theory in a Spinozian and Marxist vein. I would note, however, that Lordon's insistence on analyzing the *conatus* in the world-system's social relations after the Industrial Revolution ignores the mechanisms to achieve consent or domination among new subjects in the Spanish-held Americas. This, despite his acute observation that Spinoza's *Ethics* are informed by the financiers' society in the New Provinces where the author lived, themselves dependent on the bullion coming from America and their interest and profits made from loans and investments in New World ventures.

11. Per Bourdieu (1993, 190), "The source of historical action, that of the artist, the scientist, or the member of government just as much as that of the worker or the petty civil servant, is not an active subject confronting society as if that society were an object constituted externally. The source resides neither in consciousness nor in things but in the relationship between two stages of the social, that is, between the history objectified in things, in the form of institutions, and in the history incarnated in bodies, in the form of that system of enduring dispositions which I call habitus." My conjunction of the two terms brings the question of subjectivity to the fore without resolving it. Clearly, as a figure, metalepsis insinuates a relation between the producer and the production, though its truth proposition is not so much represented as performed.

12. For the historical significance of messianism experienced as pachakuti in the 1550s and 1560s, see MacCormack 1988; Wachtel 1973; Galindo 1986.

13. For a cultural history of the cannibal trope, see Carlos Jáuregui, *Canibalia* (2005, 2008).

14. Currently, this is the case of Shariah law in Africa, which is treated as jus gentium within the international paradigm of human rights law, even though Shariah itself makes claims to universal jurisdiction.

15. An encounter similarly recounted in López de Gómara 1987 and Cervantes de Salazar 1914.

16. Known by its shortened Latin title, *De unico modo*.

17. For a time line, see Schüssler 2005.

CHAPTER ONE   On the Same Boat

The epigraph from *Diccionario de Autoridades* reads, "El mucho desorden trahe orden. Refrán con que se da a entender que los gastos supérfluos y prodigalidad acarréan pobréza y miséria: y ella oblige à la moderación y buen gobierno."

1. See Carl Schmitt's *The Nomos of the Earth* (2003) for his now-classic discussion of the violent origins of the law in the delimitations of land and Spanish imperialism in the early modern period. See the special issue, "Reorienting Schmitt's Nomos," in *Política común*, edited by John D. Blanco and Ivonne del Valle, for inquiries into the new forms of subjectivity that arose from those areas "beyond the pale" (the colonized on either side of the Tordesillas line) in a global, yet demarcated, spatialization of primitive accumulation, foundational violence, and Catholic evangelization.

2. However, see Bauer 2003 for his reading of the *Ordenanzas* that reinforces their premises for proper functioning.

3. See Angel Rama's *The Lettered City* (1996) for an account of Spanish Empire that follows these parameters. As with Todorov, the *Ordenanzas* inspired Rama's vision of America emerging "Athena-like" from Spain's utopian vision of empire. In contrast, Baber's (2010) account of Tlaxcala elites negotiating the legal recognition of Tlaxcala as a city in the early to mid-sixteenth century points to native contributions in the development of Spain's imperial bureaucracy. See also Rappaport and Cummins 2012 for Indigenous writing "beyond the lettered city" in the Andes.

4. For François Perroux (1948, 18), Schumpeter's vision of the entrepreneur, which confounds abstraction with suggestion, exhibits "an epic sublimation of modern enterprise." In contrast to the production of *knowledge* of capitalism, of which the entrepreneur's "foreknowledge" is but a subset, Philippe Pignarre and Isabelle Stengers (2005, 66–71) explore the possibilities of *yearning* for liberation that break the knowledge/belief binary.

5. *Indios encomendados* were the Indigenous subjects who provided tribute in labor and in goods to the holders of encomienda in exchange for Christian tutelage. For more on the encomienda system see Arranz Márquez 1991; Mira Caballos 1997.

6. In this way, his proposals on the relationship between trading violence, sovereignty, debt, and political power foreshadow the concerns of contemporary scholars and activists such as David Graeber (2014) and Jason W. Moore (2016), though Acosta was advocating *for* imperial expansion and consolidation, not against it.

7. The terms are outlined in chapter 103 of the unattributed *Historia compostelana* (Anonymous).

8. See D'Arienzo (2003) for reconstructed maps of Lisbon in the thirteenth, fourteenth, and fifteenth centuries. The city's distribution is begging for a reading in terms of "strategy" and "tactics" along the lines of Michel de Certeau's influential "Walking in the City" in *The Practice of Everyday Life* (1984). For more on Genovese communities in the Mediterranean that required similar concessions in other cities, like Seville or Soldaia in the Ukraine, in return for managing their navies, see Day 1988; Poleggi 2008; Walton 2015.

9. Francis Bacon (1561–1626) writes that "all states that are liberal toward naturalization are unfit for empire," though Spain's empire offered an exception worthy of note to the English statesman and essayist: by "employ[ing] almost indifferently all nations in their militia of ordinary soldiers, yea, and sometimes in their highest commands" Spain was able "to clasp and contain so large dominions with so few natural Spaniards" (Bacon 1999, 150–51).

10. Soon after the crown's cessation of payments in 1597, the Castilian monarchy followed the Portuguese monarchy's example in issuing *asientos de esclavos* (slave contracts) as repayment for capital loans to European financiers as a matter of law (Sanz Ayán 2004, 36–37). However, as a matter of fact, the capitulación between Ferdinand of Aragon and Pedrarias Dávila shows an instance of giving an asiento de esclavos as a method to finance the conquista in Tierra Firme in the early sixteenth century (Morales Padrón 2008, 90–91). These items on Indigenous and African enslavement in the contract are analyzed further for their implications for our understanding of financing conquest in the next chapter.

11. In 1517, the crown assumed control of refining and minting gold for the House of Coin. However, Gaspar Centurione held and sold the Mexican gold at a public auction before refining, minting, and liquidating the fund (and releasing of the quinta real) could proceed. By 1522, Stefano Centurione was running the public auctions of American gold for the House of Coin.

12. For analysis of the first series of funds and their expeditions in the Americas, see Otte 1978, 2008.

13. The misfortunes of the Grimaldi family as a whole were short-lived as evidenced by the *letras* (loans) signed between Charles I and later Philip II.

14. Sanz Ayán (2004, 1–20) offers a comprehensive panorama of the complex family and "national" networks involved in trade, finance, and politics in Castile and Aragon under the regency of Ferdinand II of Aragon and Juana I of

Castile. For an in-depth analysis of the negotiations behind the Spanish crown's various bankruptcies, see Drelichman and Voth 2014.

15. Following Philip II's decision to cease payments to creditors in 1597, the Medio General, a consortium of creditors made up mostly of Genovese bankers, justified its right to charge and receive back interest on loans based on moral arguments that referred to the price of peril (Sanz Ayán 2004, 28). However, suffering their debtor's default on a loan would exemplify the instance of peril that gave moral legitimacy to the creditors' moneylending in the first place.

16. As in the capitulaciones signed with Diego Colón and Pedrarias Dávila, among others.

17. See Elvira Vilches's (2010) analysis of the confusion of value, specie, and form generated by the influx of American bullion in Spain during the sixteenth and seventeenth centuries.

18. Cotruglio Raugeo (1416–69), or Kotruljevic of Ragusa, is thought to have written the book around 1400. The title page of the Venetian edition that was published in 1573 notes, "scritti da piu di anni cx et hora data in luce" (written over more than one hundred ten years ago and now brought to light).

19. Covarrubias (1611, 492) makes the equivalence between the trustworthiness of a person's word, especially a merchant's, and a person's wealth, as someone who always pays his debts: "la credulidad que damos a lo que se nos dize. Credito, buena opinion y reputación. Credito entre mercaderes, abono de caudal, y correspondencia con los demas. Acredita a uno abonarlo. Acreditarse, cobrar credito. Acreditado, abonado. Desacreditar. Desacreditarse" (the belief with which we accept what is said to us. Credit, good opinion and reputation. Credit among bankers, payment in installments, and exchange with others. Discredit. Discredit yourself). See Sprague's *Romance of Credit* (1943) for a spirited endorsement of the capitalist's word as his "credit" just as the term "venture capital" was coming into vogue. Sullivan's *Rhetoric of Credit* (2002) similarly references double-entry bookkeeping and merchants' diaries to emphasize the interpersonal exchanges of capitalism in Jacobean London and speculate on the reception of plays that represented city exchanges. Sullivan's recourse to these merchants' manuals underscores the humanism of their endeavours in an effort to contest Agnew's *Worlds Apart* (1986), a study of Jacobean plays that largely emphasizes the alienating effect of commercial discourse on the majority of audiences in seventeenth-century London.

20. In the parlance of contemporary venture capitalists, a contrast is made between "capital equity" and "sweat equity," usually with time and labor receiving greater weight in the distribution of profits at the entrepreneurial level.

21. In the *Politics*, Aristotle makes a distinction between commerce (which includes seafaring), usury, and labor. Yet all forms of wealth procurement beyond household management are the object of the Philosopher's derision:

There are two sorts of wealth-getting, as I have said; one is a part of household management, the other is retail trade: the former necessary and honorable, while that which consists in exchange is justly censured; for it is unnatural, and a mode by which men gain from one another. The most hated sort, and with the greatest reason, is usury, which makes a gain out of money itself, and not from the natural object of it. For money was intended to be used in exchange, but not to increase at interest. And this term interest, which means the birth of money from money, is applied to the breeding of money because the offspring resembles the parent. Wherefore of all modes of getting wealth this is the most unnatural. (I.x)

It should be noted here that in *The Kingdom and the Glory* (2011), Agamben's archaeology of the political theological economy of grace is circumscribed, in part, by this Aristotelian definition of economy, which, like the law (nomos), notoriously does not apply to the sea and its ventures.

22. As outlined by the Real Cédula of 1503 that ordered the creation of the Casa de Contratación. Keeping capital investment at a minimum also reflected the constant threat of the crown's impending insolvency. The crown's insolvency culminated in crises of 1575 and 1597, when the crown suspended payments on principal and interest of loans from Genovese, Austrian, and Castilian bankers.

23. "Capitalism" as a term to describe the economic activities of Europe's elite during the early modern period has been met with resistance. The compunction to close the door on the term responds to the inherent circular logic behind comparisons, especially across temporal divides. Yet as Braudel (1977, 45) concedes, "Certain mechanisms occurring between the fifteenth and eighteenth centuries are crying out for a name all their own. When we look at them closely, we see that fitting them into a slot in the ordinary market economy would be almost absurd. One word comes spontaneously to mind: *capitalism*. Irritated, one shoos it out the door, and almost immediately, it climbs in through the window. Capital denotes not only accumulations of money but also the usable and used results of all previously accomplished work . . . but capital goods only deserve that name if they are part of the renewed process of production." In chapter 4, I show how, in his efforts to contest its legitimacy, Las Casas maps the Spanish Empire's dependence on looted grave burials in the Indies as part of the Conquest's "renewal process." For more on capitalism and periodization, see Banaji's (2010) discussion of Muslim traders' contributions to the rise of the societas in seafaring expeditions in the Mediterranean in his chapter, "Islam, the Mediterranean and the Rise of Capitalism." Banaji takes issue with historians who project Marx's analysis of industrial capitalism onto earlier forms of merchant capitalism.

24. See also Tawney's *Religion and the Rise of Capitalism* (1922) for a heroic account of Protestantism and its effects on overturning what he construes as centuries of the Catholic Church's peroration against interest and the merchant

class. Tawney's account underscores the rise of personal liberty as the result of religion assigning moral values to individual choices. For a provocative reading into capitalism's thrall and the place of Marxism in a global revolution in consciousness, see Pignarre and Stengers 2005.

25. Booty-driven, or adventure, capitalism is the term employed by Weber (2009b) to describe economic practices that rely on raids led by charismatic leaders on foreign countries for the sake of treasure (extracted from temples, tombs, mines, the chests of conquered princes, or levies on a population's jewelry or ornaments).

26. Much like the adventures of the giants Pantagruel and Gargantua and the institutions that can be found in the series written by François Rabelais (ca. 1483–1554). In *Rabelais and His World* (1968), Bakhtin reconstructs the order of the author's world by untangling the carnivalesque reversals of mores and discourse employed in the *Gargantua* series. See also Duval 1991 for a reading of the series as a humanist's rebellion against the monstrosity of Scholastic thought.

CHAPTER TWO   Contracting Love Interests

The epigraph from fray Luis de León, *De los nombres de Cristo*, reads, "Es sin duda el bien de todas las cosas universalmente la paz; y así, donde quiera que la ven la aman. Y no solo ella, mas la vista de su imagen de ella las enamora y las enciende en codicia de asemejársele, porque todo se inclina fácil y dulcemente a su bien. . . . Porque si navega el mercader y si corre los mares, es por tener paz con su codicia, que le solicita y guerrea."

1. Thus Pierre Chaunu (1991, 120) made the unequivocal distinction between *conquête* and *conquista*: "La *conquista*, non la *conquête*. . . . La *conquista* n'implique aucune action sur le sol; elle n' entraine aucun effort en profondeur pour entamer un nouveau dialogue entre l'homme et la terre. La *conquista* ne vise pas la terre, mais uniquement les hommes" (The *conquista*, not the *French conquest*. . . . The *conquista* does not imply any action on the ground; it does not bring about any effort in depth to start a new dialogue between man and the land. The *conquista* does not aim at land, but only at men). In elaborating on this distinction, *conquista/French conquest*, I might add that the Indigenous subjects acted as intermediaries between the conquistadors and the "new" lands through systems of indirect rule.

2. Seed (1995) coined the term "ceremonies of possession" to explore the differences between the performative though conflicting modes of empire making through different rituals of taking among the Spanish, English, and French. Thus, for the Iberians, "first" sightings of lands combined with Adamic-naming gestures (100–148) gain new potency with the *Requerimiento* and its scripted performance of new subject and Christian neophyte making (69–99).

The English, on the other hand, privilege the possession of territory, hence clearing and staking of land and the creation of a frontier creates "new" territories and categories of belonging on the spatial demarcation between those within and without the border fences (16–40). Finally, the French stress alliances with Indigenous peoples by punctuating areas for trade with stone markers, also denoting territorial possession (41–68).

3. In Greek, *nomos* is a unit of land. Schmitt (2003, 70) analyzes nomos as "the *measure* by which the land in a particular order is divided and situated; it is also the form of political, social and religious order determined by this process. Here, measure, order, and form constitute a spatially concrete unity" (original emphasis).

4. The so-called Papal Donation refers to the three bulls promulgated by Alexander VI in 1493. *Eximiae devotionis* (May 3), *Inter caetera* (May 4), and *Dudum siquidem* (September 26) sought to incorporate "discoveries" made by expeditions managed by the Spanish monarchs into previous schemata for conquests of terrae incognitae developed with Portuguese sovereigns in the fifteenth century. The bulls set the stage for the negotiations that led to the Treaty of Tordesillas in 1494. What exactly the pope had donated and whether he had the right to do so were questions hotly debated by Fernández de Enciso, Palacios Rubio, Francisco de Vitoria, Bartolomé de Las Casas, Juan Ginés de Sepúlveda, and Francisco Suárez, to mention a few notable authors.

5. Taking her cue from Las Casas, Seed argues that the *Requerimiento* originates in Muslim practices of conquest. See Derrida's "On Absolute Hostility" in his *Politics of Friendship* (2005) for a sustained analysis of the contradictions inherent to Aquinas's distinction between *inimicus* and *hostis*.

6. See O'Gorman's *La invención de América* (2006) and Rabasa's *Inventing America* (1993).

7. Sorcery! See Pignarre and Stengers 2005.

8. In *De unico vocationis modo* (1990), Las Casas argued that the Christian neophyte's consent had to be freely given for an authentic conversion to occur. He nonetheless recognized that conversion was a painful process.

9. Ranajit Guha provides an indispensable framework for reading colonial documents as counterinsurgent texts in *Elementary Aspects of Peasant Insurgency in Colonial India* (1983), which then supply indexes of subaltern subjectivity and ingenuity in their acts of resistance against empire. Guha's view of imperial listening (through filters, as it were) can be contrasted with Gayatri Spivak's "Can the Subaltern Speak?" (see Spivak 1996) and Rabasa's (2010, 282) rejoinder, can we listen to and enter in dialogue with the subaltern?

10. As explored by the figure of Socrates in Plato's *Symposium*, eros was a painful passion, similar to allegory in its effect on the subject. See also De Man (1979) and Roilos and Yatromanolakis (2003) for their interpretations of allegory in Platonic terms. The Hellenist vision of eros, more generally, is a passion

induced by the gods in humans; i.e., it is of an external origin to humanity. *Storge*, the love of family, is a duty that, for Aristotle, edifies the commercial pursuits of merchants. The duty to provide for a loved one elevates an occupation otherwise contemptible for its pursuit of material gains. *Agape*, as employed in the New Testament, refers to altruistic love. *Philia* refers to brotherly love. Both *philia* and *agape* comprise the Latin usage of *caritas*, especially as employed in Aquinas.

11. "Virtue," when accounting for its Latin etymology, is a highly gendered term. Virtue, from *virtu* < *vir*, meant "manliness" in classical Latin. For discussions of Machiavelli's constant references to *virtù* as a discursive effort to eschew Christian ideas of virtue in favor of the original Latin sense of "manliness" or "prowess," see Price 1970; Nederman 2000; Wood 1967.

12. Numerous studies of lyric and love poetry could be mentioned here. I am particularly fond of María Rosa Menocal's *Shards of Love* (1994).

13. See Curtius 1990 for a classic study of this trope. See also Navarrete 1994 for the expression of this trope in the poetry of golden age Spain in the aptly titled *Orphans of Petrarch*.

14. By no means an exhaustive list, these are some of the most representative works on caritas written in Spanish in the sixteenth century: *De los nombres de Cristo* (1583), by Luis de León; *Tratado de la vanidad del mundo* (1574), by Diego de Estella; *Llama de amor viva* (ca. 1585–86), by Juan de la Cruz; and *Moradas* (1577), by Teresa de Jesús.

15. Ercilla's diatribe against interest in *La Araucana* (1569–89/90) is a classic example of the depiction of private interest as the public good's (*bien público*) foremost enemy (Cantos III and XXV). In one of the more brazen efforts to reconcile caritas and cupiditas, Balbuena's *Grandeza mexicana* (1604) offers a celebration of Mexican mercantile capitalism that would place private interest at the service of the Christian, public good and vice versa. This is not to say that Ercilla's rejection of *amor* as "love interest" in Cantos III and XXV indicated an overall rejection of the imperial project. Ercilla was certainly critical of the empire he served. So, too, were Fernández de Oviedo and Acosta, whose appraisals of interest are analyzed in chapters 3 and 4, respectively.

16. As in Petrarch's infamous confession that his covetousness for Laura is a form of idolatry. The last three lines of the sestina read as follows: "L'auro e i topacii al sol sopra la neve / *vincon* le bionde chiome presse a gli occhi / che menan gli anni miei sì tosto a riva" (gold and topaz in the sun above the snow / *vanquish* (or *are vanquished by*) the golden locks next to her eyes / that lead my years so quickly to shore" (Petrarch 1976, 89). These lines overturn the hierarchy of caritas over cupiditas in Psalm 118 of the Vulgate: "ideo dilexi mandata tua super aurum et topazion" (I have loved thy commandments above gold and topaz).

17. Ollman's (2010) inquiry into the alienation of the capitalist entrepreneur informs my own understanding of Charles V's moral economy as codified in the Laws of the Indies. An orthodox Marxist in the Department of Politics at New York University, Ollman started a company to sell a board game he called Class Struggle. This experimentation with his own consciousness led to a compelling argument on the alienation of the capitalist.

18. Note that Vas Mingo and Morales Padrón, editors of the capitulaciones, leyes, and ordenanzas consulted for this chapter, dispute the category "contract" for the documents that set out the obligations of crown and crew to each other. They prefer to analyze the capitulaciones in terms of medieval suzerainty, of services and gifts exchanged between a liege lord and his loyal subjects. However, their distinction between feudal and mercantile relationships, as we saw in the cases of Portuguese-Genovese admirals in chapter 1, is overstated. My approach to the capitulaciones is shared by other scholars who have analyzed the financing of conquests in the Atlantic world such as Giovanna Montenegro (2017) and Anna More (2014).

19. The Treaty of Tordesillas defines the meridian as 370 miles west of the Cape Verde Islands. Subsequent Spanish and Portuguese navigators, including Fernández de Enciso, who brings up the stern in Guamán Poma's allegory of conquest, attempted to define the demarcation line in degrees.

20. I use *magical* with reference to Mauss's problematic distinction between religion and magic in his *General Theory of Magic* (2001). See also Malinowski's *Magic, Science and Religion* (1992) and Tambiah's *Magic, Science and Religion and the Scope of Rationality* (1990). All three authors attempt to define magic by making either an explicit or an implicit contrast with science and religion and secret and public practices. Yet such distinctions are themselves rooted in church doctrine.

21. Las Casas (2005, 43) later condemned the document with emotional appeals to truth, justice, and Christianity. See Hanke (1938) and Muldoon (1980) for an apologetics of the *Requerimiento*. However, readers may also wish to consult Gutiérrez (1993, 113–25) for his response to charges of anachronism leveled against critics of the *Requerimiento*.

22. On the holding pens for Indigenous captives, Pedro Romero gave eyewitness testimony in the *Juicio de residencia a los jueces de Apelación* (1516) (quoted in Mira Caballos 1997, 278). For a classical definition of irony, see Quintilian on allegory (VIII, vi). For the rhetorician, the force of both tropes, irony and allegory, resides precisely in their literal interpretations.

23. Las Casas traces the origin of the encomienda system to a misreading of a letter written by Isabel of Castile days before her death. The letter itself, reproduced by Las Casas in chapter 14 of the third book of his *Historia de las Indias* is discussed at greater length in chapter 4 below. Andre Saint-Lu omits Isabel's letter in his edition of the *Historia de las Indias*, but it is included in

earlier, nineteenth-century editions of the work, as well as in the volumes dedicated to the *Historia* in the edition of *Obras completas* under the direction of Castañeda Delgado and Huerga.

24. Much more could be said about the regulation and demarcation of time for the purposes of venture capitalism in these laws and those of the 1526 *Ordenanzas*, as well as the New Laws of 1542. For a compelling account of the homogenizing effects of industrial capitalism in England, see Thirst 1996. He argues that workers replaced, that is, destroyed, their own consciousness of time—seasonal labor, social events, tasks—with the owners' time, a "future oriented calculative rationality" (567). See also Kuriyama's "The Enigma of 'Time Is Money'" (2002) for an engaging reflection on the relationship between capital, time, and consciousness in a nonmodern society, Meiji Japan, in particular, in the process of learning Western ways.

25. To the extent that the purpose of bureaucracy is to ensure unity of action by limiting opportunities for actual decision making (Weber 2009a, 196–224).

26. These *items* include recommendations to avoid making promises to the Indians that cannot be kept, injunctions against gambling, regulations on inheritance, and the grains to be cultivated in Golden Castile.

27. The construction of a palisade in the public square recalls what, in *The Porcelain Workshop*, Negri defines as that "place where the individual can distribute gifts to friends and inflict death on enemies" (2008, 17).

CHAPTER THREE   Telling Islands in the *Claribalte* and the
Historia de las Indias

1. My concern is not to trace the origins of the modern Latin American novel from the primary resources afforded by the legal conventions of the colonial period and the creation of archival fictions (González Echevarría 1998) or to reconstruct the convention for creating fictitious places in modern Latin American narrative (e.g., Santa María, Macondo, Comala) out of the "lost," Indigenous cities forever on the horizon in colonial texts (Adorno 2008). Instead, I trace narrative *modes* in sixteenth-century texts that grapple with representing or, alternatively, repressing traumatic or shameful events and the authors' roles in *producing* them as letrados, more along the lines of Hayden White's treatment of the ethics of writing history in *The Practical Past* (2014).

2. Vargas Llosa infamously characterized Hispanic colonial letters as one bereft of fiction—a reflection, for the famed Peruvian "boom" author, of a society whose creative inclinations were supposedly stymied by the Inquisition's widespread censorship. See Adorno's (1992b) introduction to Leonard's *Books of the Brave* for a thorough exposition of the contradictions and misrepresentations in Vargas Llosa's vision of colonial letters and society. As for my use of the

term "ethnographic" to describe Fernández de Oviedo's major works, which is not included as a genre in Mignolo's now-canonical essay "Cartas, crónicas, y relaciones del descubrimiento y la conquista," I contend that it is a colonial mode of writing employed across various of the recognized genres that comprise the colonial corpus and foundational to the "meta-historiography" of colonial texts describing indigeneity.

3. More precisely, Gerbi (1949, 378) writes of the *Claribalte*, "La novela caballeresca de don Claribalte es una de aquellas obras irritantes que plantean más problemas, provocan más expectativas y dejan a uno con más dudas de lo que estaría justificado por su valor intrínseco" (The chivalric fiction *Don Claribalte* is one of those irritating works that pose more problems, generate more expectations, and leave one with more doubts than what would be ordinarily justified by their intrinsic worth).

4. See Reséndez (2016, 63) for examples of the brands used to distinguish Indigenous slaves of "war" (*de guerra*) from Indigenous "ransomed" (*de rescate*) slaves. The contract between Fernando II of Aragon and Dávila gives the latter a license to capture Indigenous slaves by war. "Ransomed slaves" involved a transfer of captive laborers from their "Indigenous owners" to the Spanish (see Moscoso 1986). Rescates were justified as an extension of pre-Hispanic Indigenous practices of captivity. Proponents of rescates' legality argued that these slaves would see their lots improved, for as Spanish captives they would be exposed to Christianity. For histories of Indigenous slavery in the Spanish-held Americas prior to 1542, when the New Laws categorically prohibited Indigenous slavery, and the continued practice of Indigenous enslavement after 1542 despite these prohibitions, see Reséndez 2016; Deive 1995; Saco 1932.

5. For a discussion of scholarly debates on the possible influence of the Pedro Serrano anecdote of Inca Garcilaso's *Comentarios reales* I (1609) on Daniel Defoe, see Voigt 2009, 91–98.

6. Petrarch describes the ceremony, which he claims to have witnessed, in *De vita solitaria* (II.xi). As D'Arienzo contends, the titles to the Canary Islands may have been rewarded to the Castilian claimant following a failed Florentine-Portuguese expedition to the same islands in 1341. Boccaccio gives an account of this voyage in *De Canaria et de insulis ultra Hispaniam in Oceano noviter repertis* (1964).

7. See Greer, Mignolo, and Quilligan 2007 for the origins and continued relevance of the Spanish black legend.

8. For more on the Fuggers in Venezuela, see Montenegro 2017; Johnson 2008.

9. Rascón and Quintero were co-owners of the *Pinta*, but Martín Alonso Pinzón was the ship's captain.

10. As Nicolás Wey-Gómez (2008) has recently reminded us, early modern Europeans subscribed to ideas about the tropics that linked this geographic latitude with inhabitants' inferior character traits (i.e., their aptitude for "natural

slavery") and an abundance of natural riches (gold, wood, etc.). Columbus's "southing" (on a westerly course) points to his quest for *natural* riches both in slave labor and in material resources.

11. The encounters between European, American, and Asian civilizations during the sixteenth century, as explored by Serge Gruzinski in *The Eagle and the Dragon* (2014), follow the patterns of conquest—on the European side—that emerged out of the mid-Atlantic islands and the West African laboratory for imperialism and that Las Casas analyzes in great detail in the *Historia de las Indias*.

12. The dichotomies created by the rhetoric of Las Casas in the *Brevísima relación* are superficial and mainly aligned with the epithets he employs to distinguish between "indios" (*mansos, ovejas, generosos, buenísimos*, etc.) and "los españoles" (*crueles, lobos, codiciosos, malísimos*, etc.). At the same time, the leading role given to the voices of Indigenous insurgents (famously, Hatüey) undercuts the text's emphasis on Indigenous docility. Instead, in the section on New Spain we are given the dictum that "ninguno es ni puede ser llamado rebelde si primero no es súbdito" (nobody is or should be called a rebel if he is not first a subject) (112). The chapter on the conquest of Venezuela similarly complicates the dichotomies established in the opening sections of the text. Rather than Spanish conquistadors, nominally Catholic according to Las Casas, we have German conquistadors, possibly Lutheran bankers (116–17); his narrative of the martyrdom of the Dominican missionaries posits an aporia where the missionaries received an unjust martyrdom at the hands of Indigenous insurgents who were, nonetheless, defending themselves justly against Spanish invaders (109–10). For more on the German conquistadors and bankers in Venezuela, such as Nikolaus Federmann, see Christine Johnson's *German Discovery of the New World* (2008) and Giovanna Montenegro's "Conquistadors and Indians 'Fail' at Gift Exchange" (2017).

13. In *Marvelous Possessions*, Stephen Greenblatt argued that in his *Carta a Luis de Santangel* (1493) Columbus invented the conjunction of the "most resonant legal ritual he can summon up" (i.e., taking possession of the Indies) with the most resonant emotion (i.e., wonder). He explains:

> By itself a sense of the marvelous cannot confer title; on the contrary, it is associated with longing, and you long precisely for what you do not have. . . . But something happens to the discourse of the marvelous when it is linked to the discourse of the law: the inadequacy of the legal ritual to confer title and the incapacity of the marvelous to confer possession cancel each other out, and both the claim and the emotion are intensified by the conjunction. Neither discourse is freestanding and autonomous; on the contrary, each—like individual words themselves—takes its meaning from its conjunction with other motifs, tropes, and speech acts, and from the situation in which it is inserted. And there is a further motive for the conjunction: under the actual circumstances of the first encounter, there was no discourse adequate

for the occasion. In the unprecedented, volatile state of emergence and emergency in which Columbus finds himself, anything he says or does will be defective. His response is to conjoin the most resonant legal ritual he can summon up with the most resonant emotion. (Greenblatt 1991, 81)

In the *Brevísima relación*, by referring to the "marvels" wrought by the destruction of the Indies by the conquest, Las Casas zeroes in on the legal and emotional conjunction created by Columbus's attempt to unravel the potent conjunction of terms.

14. For more on the discursive violence of this picturesque scene, see Rabasa 1993.

15. Despite the name, the "long sixteenth century" per Braudel and Moore begins around 1450 with the Iberian monarchs' co-ventures with Italian merchants, seafarers, financiers, and admirals of national fleets. Note that these personae could often be embodied by the same person, as with Bartolomé Perestrello whose case is the subject of this chapter.

16. I cannot help but see Perestrello and Columbus as prefiguring the slaveholder of the antebellum South, described by Johnson (2013, 208) as a figure whose eschatology was "rooted in his ecology [. . . as] human beings, animals and plants [were] forcibly reduced to limited aspects of themselves, and then deployed in concert to further slaveholding dominion."

17. As to the providential nature of the discovery, Las Casas added that it was impossible to know God's intent because "God's judgments are profound and no human can or should penetrate them" (*los juicios de Dios son profundísimos y ninguno de los hombres los puede ni debe penetrar*) (Las Casas 1994, 560).

CHAPTER FOUR   The Specter of Las Casas in the Political Theology
of José de Acosta

1. See Conley 1992 and Rabasa 2000 for the reception and translation of the *Brevísima* in Protestant Europe. The *Brevísima* was the only work by Las Casas that was translated and printed in modern European languages during the sixteenth century. Manuscripts of *Historia de las Indias*, *De unico vocationis modo*, *Apologética historia sumaria*, *De regia potestate*, *De Thesauris*, etc., were circulated and read by members of his activist network in Iberia and the Indies. See Parish 1967 for detailed accounts of the circulation, reception, and publication of Las Casas's early works. See Losada 1992 and Denglos 1992 for the circulation and reception of *De Thesauris* and the *Doce dudas* (in *Obras completas* [Las Casas 1988–]). Acosta's *Historia natural y moral de las Indias*, in contrast, was an instant "best seller" in Spain, Italy, France, and England, both in translations into modern languages and in Latin. See also Del Valle 2009 for her discussion of the influence of the *Spiritual Exercises* by the founder of the Society of Jesus, Ignatius of Loyola (1491–1556), on Acosta's *Historia natural y moral de*

las Indias. To place Acosta's oeuvre in the longue durée of Spanish scientific writings and imperialism, see Cañizares-Esguerra 2006a. A little over a hundred years after the first publication of the *Historia natural*, Giambattista Vico (1668–1744) referenced Acosta and Francisco Suárez (1548–1617) in his exposition of the poetic history of primitive peoples in the *New Science* (1701).

2. For comparisons to liberation theology, see Gustavo Gutiérrez's *Las Casas: In Search of the Poor of Jesus Christ* (1993) and Walter Mignolo's introduction to the English translation of Acosta's *Historia natural y moral de las Indias* (xxi). For the Lascasian origins of "teología india," see Rabasa's *Tell Me the Story of How I Conquered You* (2011). In defense of his use of the term "teología india" against the inquiries of the Office of the Doctrine of the Faith, López Hernández reminded church authorities that it was Bartolomé de Las Casas who originally coined the term *theologia indorum*, or theology of the Indians.

3. In the spirit of Michel de Certeau's (1988, 2) paradoxical assertion that "the ghosts find access through *writing* on the condition that they remain *forever silent*" (original emphasis).

4. The posthumous publication in Frankfurt of *De regia potestate* (1571), which argued in favor of autonomous native rule, has led researchers to question its authorship. In his introductory study, Pérez Luño (1990) points to evidence in favor of Las Casas's authorship.

5. For the larger context of the Lima Inquisition in the empire as a whole, see Lea 2011. The proceedings against Francisco de la Cruz can be read in the editions by Abril Castelló (1996) and Abril Castelló and Abril Stoffels (1996). See Bataillon 1996 for a persuasive argument that the proceedings against Francisco de la Cruz had wider repercussions for Las Casas and his followers in Peru.

6. See Jaúregui 2009 for a discussion of how Indigenous anthropophagy was received, interpreted, and connected to the Eucharist in the sixteenth and seventeenth centuries.

7. See Lisi 1990; Trujillo Mena 1981. In 1588 Acosta personally brought the minutes of the proceedings and decisions of the Third Council of Lima to Rome for the pontiff's approval. Philip II ratified the minutes in the Escorial in 1591.

8. As demonstrated in Horswell 2005, the Spanish tradition of material and discursive persecution of "efeminados" and "sodomitas" was employed in Peru to justify the conquest, once colonists had experienced and observed third-gender rituals. Horswell's *Decolonizing the Sodomite* offers a history of Indigenous gender and sexuality in Peru by taking Ranajit Guha's *Elementary Aspects of Peasant Insurgency* (1999) as his methodological starting point.

9. In the *Brevísima* Las Casas shows his appreciation for cultivated fields and derision for the Spanish rush for gold. It is not that Las Casas disparaged all means of *usufruct*, rather he valued sweat and toil that did not get subsumed into the processes of capital. Unlike Acosta, Las Casas (2005, 84) did not count the activities of the conquistadors as labor.

10. O'Gorman, in the introduction to his edition of Las Casas's *Apologética historia sumaria* (1967, clxvii).

11. Las Casas (1994) refers to him derisively as the *"Bachiller* Anciso" who believed in the legal fictions of the *Requerimiento* because he was an argumentative lawyer: "como Anciso era jurista, debió parecerle que justificaba, con usar del requerimiento, mejor sus robos y violencias que iba a hacer a los vecinos de Cenú" (as Anciso was a lawyer, he must have thought that making use of the *Requerimiento* justified the theft and violence he planned against the residents of Cenú) (1994, 2019).

12. In "Security, Territory and Population" (1997) Foucault tied "massifying" forms of biopower to the emergence of the absolutist European state. According to Hardt and Negri (2000, 24), "Biopower is a form of power that regulates social life from its interior, following it, interpreting it, absorbing it—every individual embraces and reactivates this power of his or her own accord. Its primary task is to administer life. Biopower thus refers to a situation in which what is directly at stake in power is the production and reproduction of life itself." Their example of the limits of empire invokes the self-immolation of Buddhist monks in Tibet.

13. See Pagden's *The Fall of Natural Man* (1982).

14. The longevity of these titles, just or not, continues to manifest itself. Consider the "defense of the innocent" and "freedom from tyranny" arguments invoked by President Barack Obama before air strikes on Libya in 2011.

15. We—who is this "we"?—might refer to them as "crimes against humanity" today.

16. More precisely, if Vitoria had made references to cases from the conquistas in the Americas during his lectures in the 1530s, his students made no note of them.

17. There are more recent figures of trade at gunpoint, for example, Commodore Matthew Perry's turning of his ship's cannon on Edo to open up Tokugawa Japan to American goods.

18. Kant will make similar claims in *On the Perpetual Peace* (2015).

19. Las Casas concludes by referring to chapters 19, 22, 24, and 25 of the first book of his *Historia de las Indias* to remind readers that Portugal initiated the process of "free" trade and evangelization in Africa (Guinea), which was later followed by Castile.

20. Las Casas paraphrases Christ as follows:

Christus etiam prohibuit Evangelii sui promulgatoribus ne possiderent aurum vel argentum nec pecuniam; multo fortius ne ab his, quibus praedicaturi erant, non solum ut non violenter raperent aut ab invitis tollerent, verum etiam nec ab volentibus libenter dare acciperent. (*De unico vocationis modo* 416)

[Christ prohibited the preachers of the Gospel from carrying gold, silver, or money; and not only were they not to rob the men whom they came to preach, or take anything against their will, but also, they were not to accept any thing that [the Gentiles] would willingly give them.]

21. I have changed McIntosh's translation of *omnes Indorum Satrapas*, "men and Satraps of the Indies," to more accurately reflect the terms used by Acosta to describe Indigenous forms of self-government. Acosta's use of the generic "men" (*omnes*) for the entrepreneurs of conquest and the alienating term "satraps," lords of provinces in the Persian Empire, for Indians leaves little doubt as to his views on Indigenous sovereignty in the context of Christian empire.

22. Many are the authors who have spilled ink on the thorny question of exclusion in Cyprian's phrasing. A more recent consensus, shared by Cardinal Ratzinger (later, Benedict XVI) (1963), González Ruiz (1962), and Gutiérrez (1993), emphasizes the unity of the church as a route to salvation. However, based on Cyprian's phrase, writers in the Augustinian tradition made a stark delineation between Christians and non-Christians. For an understanding of Thomistic thought on universal salvation, see Capéran's classic *Le problème du salut des infidèles* (1912).

23. In *De veritate*, Aquinas posits the problem of the savage child: if the child is raised among barbarians but during the *time* of the law of grace, is the child shut out from God's salvific grace? For Aquinas, if the child follows natural reason in search of the good, on attaining the age of reason, God, by some extraordinary measure, would intervene (in *Opera omnia* 14.11.1). But how? And if God would use extraordinary measures for one savage child, what of a multitude?

24. Sepúlveda's most vehement arguments in favor of violent evangelization were articulated in his *Apologia*, which is earlier than the *Democrates alter*, his best-known work. In the *Historia general y natural de las Indias*, Fernández de Oviedo uses even more demeaning language to deny rational thought to Indigenous peoples.

25. Both jurists belong to the modern Scholastic tradition of the School of Salamanca, which also includes Suárez. His *Disputationes* proved essential to the thought of Giambattista Vico.

26. See Dussel's "Alterity and Modernity" (2007) for a reading of Las Casas's entire body of work as squarely at odds with modernity and European imperialism. For an opposing viewpoint, based almost entirely on the early work of Las Casas, see Castro's *Another Face of Imperialism* (2007).

27. The contingency of benevolence on the acceptance of Western-held universal truth claims continues to privilege the refusal of one party to dialogue. Consider, as an example, the liberal position of John Rawls (1999, 166–67, 417), who cannot contemplate a dialogue among competing universal truth claims that does not leave benevolence at a loss on how to proceed.

28. Note that Hanke (1949, 23) attributes this line of reasoning to fray Bernardino de Mesa, King Fernando's confessor. However, the discourse of Indigenous idleness and the economy of liberty precedes by several years the Laws of Burgos, which were articulated and promulgated by Isabel of Castile in December 1503.

29. In a way that divides subjects between "old" and "new" Christians along the axis of Spanish (old) vs. native (new). This division reverberates with the old/new Christian relations on the Iberian Peninsula wherein "Spanish native" and "old Christian" are conflated against "foreigner" and "new Christian." See Martínez 2011 for the origins of racism in New Spain in Spain's "laws of blood purity" (*estatutos de limpieza de sangre*).

30. In "Ideology and Ideological State Apparatuses," Althusser (2001, 118–20) argues that ideology is constructed in the relations between what is said and what remains unsaid. The lacunae in Isabel of Castile's ideology of "liberty" are exposed when she admits to her new subjects' resistance to coercion.

31. As a reminder, the citation from *De procuranda* reads, "Itaq[ue]; necessitatis, vt dixi, cuiusdam fuit, vt suo quoadam iure, vt olim Israëliticae tribus distributore Iosue, terram sortirentur, permanente tamen, quod minimè obscurum est, supremo omnium penes Regem imperio (So, as I said, it was out of necessity, as in other times, like the tribes of Israel for example, where individuals obtained the land by lot distributed by Joseph, although as is quite clear, the supreme control of distribution always remained in the hand of the king) (Acosta 1589, 317–18). I have changed McIntosh's translation, reproduced verbatim in the first chapter, and rendered it more literally to emphasize the specificity of Acosta's allusion to Joshua 13:19.

32. "Dedique vobis terram, in qua non laborastis, et urbes quas non ædificastis, ut habitaretis in eis: vineas, et oliveta, quæ non plantastis" (Iosue 24:13, Vulgate).

33. See Gibson 1977.

34. For the argument that English colonists were similarly motivated, see Cañizares-Esguerra 2006b.

35. Viceroy Toledo's reforms (1569) predate the *Ordenanzas* and were a model for the laws promulgated by the Consejo de Indias in 1573.

CHAPTER FIVE   The Bidding of Empire

1. For the Andean archaeological record, Lanning (1967) coined the terms "horizons" to refer to the eras of interethnic political and cultural cohesion in the central Andes (i.e., Chavín, Huari, Tiahuanaco, Inca) and "periods" to refer to polities that remained "local" in their forms of organization (i.e., Paracas, Lima, Moche, Chimú).

2. See Gose 2008; Murra 2002; Duviols 1977. Though all authors agree on the pre-Hispanic importance of corn beer in Andean society, there are discrepancies on the subject of prolonged inebriation. For Gose, drunkenness and orgies emerged as a degeneration of the traditional system, caused by the stresses of conquest and the imposition of monogamy by the state and church. For a millenarian view of ritual drunkenness, see Duviols's analysis.

3. As employed by Stern (1993, 158), "mediation" and "negotiation" in the Andean context are similar to being stuck between a rock and a hard place: "The post-Incaic alliances caught native elites between traditional roles as protectors of ayllu interests, and new opportunities and demands as 'friends' of the conquistadores."

4. Note that as narrated in the *Anales de Juan Bautista*, cited by Ruíz Medrano (2010, 69), the revolt of the indios against Philip II's new tribute could *also* be interpreted as a revolt against their Indigenous overlord, Santa María Cipac.

5. For example, in 1563 Rodrigo Méndez and Pedro Avendaño rebelled in Cuzco in a bid to receive an encomienda for themselves (see *Carta del Dr. Cuenca al Rey, 30 de Abril de 1563*; quoted in Goldwert 1955, 213). In the same letter Cuenca also tells Philip II about a group of scam artists, also in the Cuzco area, who had offered to represent the interests of the curacas in exchange for 4,000 pesos. On receipt of the 4,000 pesos, the scam artist and his accomplices left their clients stranded.

6. Alonso de Villanueva and Gonzalo López obliquely alluded to the crown's lack of investment in the conquest of New Spain, noting the success of the Spanish conquest in Mexico "sin que Su Majestad hubiera gastado nada del tesoro o patrimonio real" (even though Your Majesty spent none of your treasure or royal patrimony) in the enterprise. As representatives of the encomenderos of Nueva España, they advocated a "recompensa perpetua" (perpetual compensation). Villanueva and López argued, with graphic acumen, the system's unsustainability and its oppression of the Indians of New Spain. Without perpetuity, the encomenderos, i.e., their clients, were "mercenarios y no agricultores . . . [que] sólo trataban de beber su sudor [de los indios] para luego irse" (mercenaries, not cultivators, . . . [who] would drink the [Indians'] sweat only to leave) (quoted in Hackett 1923, 1:56).

7. Suggesting that it would be even better, in good conscience, to bestow the encomiendas in perpetuity rather than sell them, Castro nevertheless conceded that a gift in perpetuity was highly unlikely. However, if the crown were to make a sale it should do so at a moderate price (Levillier 1921, 36).

8. The Consejo de Hacienda had already recommended the sale of the perpetuity of the encomienda in 1552 (Carande Thobar 1965, II:122–24). On January 26, 1556, soon after the commissioners left for Peru, Charles V abdicated the throne of Spain.

9. A detailed analysis of Philip II's counterproposal to the Consejo de Indias for granting perpetuity and civil and criminal jurisdiction *mero mixto imperi* may be found in Lohmann Villena (1977, 250) and Goldwert (1955, 353–55). As observed by Goldwert, Philip II was willing to create an aristrocracy out of the encomienda but without contiguous landholdings. Effectively, Philip II had been willing to offer Ribera more power than he had requested before the Consejo de Indias intervened.

10. See Pedro Cieza de León's *Guerra de Quito* (1553) and the *Relación de las cosas acaescidas en las alteraciones del Peru* (1553) (Las Casas Grive 2003) for contemporary accounts of the encomenderos' insurrection, led by Gonzalo Pizarro. Gonzalo Pizarro's pretensions included naming himself king of Peru, marrying a *coya* (Inca queen), granting encomiendas with jurisdiction in perpetuity, and also drafting laws to protect the Indians (Lohmann Villena 1977, 40–65). Bernal Díaz del Castillo (1984) has a colorful description of just how the rebels' requests for perpetuity were received by the members of the special junta summoned to decide the matter of perpetuity "once and for all" in Valladolid in 1550. Las Casas, Pedro de Gasca, Vasco de Quiroga, and the bishop of Michoacán also attended.

11. The uncertainty argument recalled similar reasoning by the Hieronymite fathers sent to investigate the Indians' living conditions. Their report released in 1518 recommended the perpetuity of the encomiendas as a reform effort to ameliorate the Indians' living conditions and was attacked by Las Casas as illogical and ignorant.

12. Las Casas had promoted full incorporation of indios encomendados into the crown in a letter to fray Bartolomé de Miranda in August 1556.

13. "Origo" refers to the speaker's self-defined spatial relationship to her interlocutors in the deictic markers of her utterance. As defined by Green (1992, 121–22), "deixis" refers to "the encoding in an utterance of the spatio-temporal context and subjective experience of the encoder. It is primarily linked with the speech or discourse event [that foregrounds] the encoder's subjectivity and various contextual factors . . . grammatically or lexically. . . . Any utterance refers to the speaker's 'centre' (*origo*) and surrounding cognitive environment." Personal and demonstrative pronouns, certain adverbs, time-space references, vocative particles, and subject modifiers (adjectives and past participles that generally decline along gender, feminine marked) are deictic terms. However, verbs conjugated in the first person are also an obvious marker of the "zero-point," while discourse organizers are more obscure indicators. Green emphasizes that whether an utterance is deictic or not depends on the speaker's and interlocutor's shared context. Thus, "deixis is distinguished by use" (123). Also, deictic terms can either be indexical or symbolic. The term "deixis," in turn, references the *epideictic* tradition in encomia and invective.

14. Might Esposito's thoughts on community and melancholy in his *Terms of the Political* (2013) be relevant to discussions of the curacas' and the friars' transatlantic gesture?

15. Stern (1993) speaks of the "tragedy of success" with reference to the Andean indios and indias who managed to thrive in the new colonial economy and make the jump from the república de indios to the república de españoles.

16. See Shell's *Children of the Earth* (1993) for his structural analysis of Christianity's universality vs. tribal origin myths. See also Marcel Detienne's *Comment être autochtone* (2003) for a nuanced discussion of the political negotiations involved in the construction of autochthony, from classical Athens to the Third Republic of France. As employed by Viceroy Toledo during the 1570s, the narrative of the original journey of the Incas from Pacaritambo to Cuzco would be used as evidence of their nonindigeneity and thus illegitimacy.

17. Sahlins's *Historical Metaphors and Mythical Realities* (1981) emphasized Indigenous forms for rationalizing contact with the Europeans on their own narrative terms. Thus the identification of Captain Cook with Lono on his first trip to the Hawaiian Islands and his killing on the second trip followed the structure of native mythology. In response, Obeyesekere in *The Apotheosis of Captain Cook* (1997) turned Sahlins's thesis on its head, arguing that the narrative of natives receiving Europeans as gods was a form of European mythology. Similar tensions have animated the scholarly debates surrounding the identification of Cortés with Quetzalcoatl and Pizarro et al. with Viracochas. Carrasco (1982), León-Portilla (2001), and Lafaye (1976), to mention but a few, accept the identification of Cortés with Quetzalcoatl in native narratives of the Spanish Conquest. Townsend's "Burying the White Gods" (2003) revises this identification in an attempt to rectify what she views as a "dehumanizing narrative" meant to satisfy European historians' needs to provide a satisfactory justification for the relatively small number of Spanish who "conquered" the Mexica and the Inca. Similarly, Adorno's (1992) treatment of Mala Cosa in the *Naufragios* by Núñez Cabeza de Vaca as a vision that responded to medieval Spanish narratives would be contested by Rabasa in *Writing Violence* (2000) with reference to shamanic practices by the Indigenous peoples of the Native American Southwest. See also Rabasa's *Tell Me the Story of How I Conquered You* (2011). I believe that the repeated gestures of scholars such as Obeyesekere and Townsend to explain away Indigenous myths as the product of a European prerogative for self-apotheosis threatens to undermine the power of myth to organize past and present events meaningfully in terms other than the rationalism of secularizing imperatives.

18. As discussed in the introduction, my use of the term "event" follows that of Badiou in *Being and Event* (2007).

19. For a summary of colonial studies' rejection of the analytical utility of the Spanish administrative arrangement of two republics for a meaningful examination of colonial culture, see Rappaport and Cummins 2012, 28–31.

20. Although Gruzinski (2002, 213) invites his readers to visualize cultural contestation not as an opposition of polar opposites but as a "series of modulations," this perspective may not prove meaningful for analyzing discrete moments of cultural confrontation. As noted by Gruzinski in *The Mestizo Mind* (2002) and also by his proponents such as Rappaport and Cummins (2012), this approach aims at an understanding of cultural contestation over the longue durée.

21. Acosta dedicates an entire chapter to fray Francisco de la Cruz's heresy in his book *De temporibus novissimis*, to illustrate the arrogance of the Antichrist (Bataillon 1966, 313).

22. Following the vague language of the earliest cédula, whose numerous loopholes may have been conditioned by Charles V's enthusiasm for the grand scale and speed of returns in the Peruvian enterprise, more precise language specifying the limits of Francisco Pizarro's authority to name a *provisional* governor, until Charles V could name a permanent replacement, was not immediately forthcoming. By 1537, however, it was apparent that Francisco Pizarro had ignored the subsequent royal cédulas when he transferred the governorship of Peru to Gonzalo Pizarro Yupangui, his son, and, in 1539, named his brother, Gonzalo Pizarro, temporary governor until his son, Gonzalo Pizarro Yupangui, would come of age.

23. "Uncanny" as explored by Freud (1977) for *heimlich*, which unlike *ocio* and *negocio* shares the same meaning with its negation, *unheimlich*.

Epilogue

1. The call for a new concept of history was made by Walter Benjamin (2007, 257) in his "Eighth thesis" in his *Theses on the Philosophy of History*: The tradition of the oppressed teaches us that the "'emergency situation' in which we live is the rule. We must arrive at a concept of history which corresponds to this. Then it will become clear that the task before us is the introduction of a real state of emergency."

2. See both articles by Sánchez Jiménez (2007, 2008) for documents held by the Archivo General de Indias that make reference to these alliances.

WORKS CITED

Abril Castelló, Vidal, ed. 1996. *Francisco de la Cruz, Inquisición, Actas I: Anatomía y biopsia del dios y del derecho judeo-cristiano-musulmán de la Conquista de América.* Madrid: Consejo Superior de Investigaciones Científicas.

Abril Castelló, Vidal, and Miguel J. Abril Stoffels, eds. 1996. *Francisco de la Cruz, Inquisición, Actas II-1.* Madrid: Consejo Superior de Investigaciones Científicas.

———. 1997. *Francisco de la Cruz, Inquisición, Actas II-2.* Madrid: Consejo Superior de Investigaciones Científicas.

Acosta, José de. 1589. *De natura noui orbis libri duo, et De promulgatione Euangelii, apud barbaros, siue De procuranda indorum salute libri sex.* Salamanca: Apud Guillelmum Foquel, Gestión del Repositorio Documental de la Universidad de Salamanca (GREDOS), Biblioteca Virtual Miguel de Cervantes. http://hdl.handle.net/10366/136856.

———. [1590] 1940. *Historia natural y moral de las Indias: En que se tratan de las cosas notables del cielo, elementos, metales, plantas y animales dellas: y los ritos, y ceremonias, leyes y gobierno de los Indios, compuesto por el P. Joseph de Acosta.* 2nd ed. Edited by Edmundo O'Gorman. Mexico City: Fondo de Cultura Económica.

———. [1589] 1996. *De procuranda indorum salute.* Translated by G. S. McIntosh. 2 vols. Tayport: C Research.

Adorno, Rolena. 1986. *Guamán Poma: Writing and Resistance in Colonial Peru.* Austin: University of Texas Press.

———. 1992a. "Cómo leer a Mala Cosa: Mitos caballerescos y amerindios en los *Naufragios* de Cabeza de Vaca." In *Crítica y descolonización: El sujeto colonial en la cultura latinoamericana*, edited by Beatriz González Stephan and Lúcia Helena Costigan, 89–107. Caracas: Biblioteca de la Academia Nacional de la Historia de Venezuela.

———. 1992b. Introduction to *Books of the Brave: Being an Account of Books and Men in the Spanish Conquest and Settlement of the Sixteenth-Century New World*, by Irving Leonard, ix–xl. Berkeley: University of California Press.

———. 2007. *The Polemics of Possession in Spanish American Narrative*. New Haven, CT: Yale University Press.

———. 2008. *De Guancane a Macondo: Estudios de literatura hispanoamericana*. Seville: Renacimiento.

Agamben, Giorgio. 1998. *Homo Sacer: Sovereign Power and Bare Life*. Translated by Daniel Heller-Roazen. Stanford, CA: Stanford University Press.

———. 2005a. *State of Exception*. Translated by Kevin Attell. Chicago: University of Chicago Press.

———. 2005b. *The Time That Remains: A Commentary on the Letter to the Romans*. Stanford, CA: Stanford University Press.

———. 2011. *The Kingdom and the Glory: For a Theological Genealogy of Economy and Government*. Translated by Lorenzo Chiesa. Stanford, CA: Stanford University Press.

Agnew, Jean-Christophe. 1986. *Worlds Apart: The Market and the Theater in Anglo-American Thought, 1550–1750*. Cambridge: Cambridge University Press.

Alfonso X, King of Castile and Leon. [ca. 1256–65] 1875. *Las Siete Partidas: Compendiadas y anotadas*. Edited by José Muro Martínez and Mariano Muro López Salgado. Valladolid: Gaviria y Zapatero.

Allen, John. 2003. *Lost Geographies of Power*. Malden, MA: Blackwell.

Althusser, Louis. [1969] 2001. "Ideology and Ideological States Apparatuses (Notes toward an Investigation)." In *Lenin and Philosophy, and Other Essays*, edited by Fredric Jameson and translated by Ben Brewster, 85–125. http://public.ebookcentral.proquest.com/choice/publicfullrecord.aspx?p=2081662.

Anghie, Antony. 1996. "Francisco de Vitoria and the Colonial Origins of International Law." *Social and Legal Studies* 5, no. 3: 321–36.

———. 2007. "The Evolution of International Law: Colonial and Postcolonial Realities." *Third World Quarterly* 27, no. 5: 739–53.

Anghiera, Peter Martyr. [1511–30] 1989. *Décadas del Nuevo Mundo*. With an introduction by Ramón Alba. Translated by Joaquín Torres Asensio and Julio Martínez Mesanza. Madrid: Ediciones Polifemo.

Anonymous. [ca. 1107–49] 1994. *Historia Compostelana*. Edited by Emma Falque Rey. Vol. 3. Madrid: Akal.

Aquinas, Thomas. [1256–74] 1996. *Thomae Aquinatis Opera Omnia cum Hypertextibus in CD-ROM*. Edited by Roberto Busa. CD-ROM.

———. [1265–74] 2010. *Summae Theologiae Secunda Secundae*. Turnhout: Brepols.

Arendt, Hannah. 1966. *The Origins of Totalitarianism.* New York: Harcourt, Brace, and World.

———. 1982. *Lectures on Kant's Political Philosophy.* Edited by Ronald Beiner. Chicago: University of Chicago Press.

Aristotle. 1999. *Nicomachean Ethics.* 2nd ed. Translated by Terence Irwin. Indianapolis: Hackett.

———. 2007. *On Rhetoric: A Theory of Civic Discourse.* 2nd ed. Translated by George Alexander Kennedy. New York: Oxford University Press.

———. 2012. *Aristotle's Poetics.* Edited by Leonardo Tarán. Translated by Dimitri Gutas. Boston: Brill.

———. 2013. *Aristotle's Politics.* 2nd ed. Translated by Carnes Lord. Chicago: University of Chicago Press.

Arranz Márquez, Luis. 1991. *Repartimientos y encomiendas en la Isla Española: El repartimiento de Albuquerque de 1514.* Madrid: Ediciones Fundación García Arévalo.

Arrighi, Giovanni. 2010. *The Long Twentieth Century: Money, Power, and the Origins of Our Times.* New York: Verso.

Auerbach, Erich. 2003. *Mimesis: The Representation of Reality in Western Literature.* Edited by Jan M. Ziolkowski. Princeton, NJ: Princeton University Press.

Augustine. [397, 426] 1995. *De Doctrina Christiana.* Edited and translated by R. P. H. Green. Oxford: Clarendon Press.

Baber, R. Jovita. 2010. "Empire, Indians, and the Negotiation for the Status of City in Tlaxcala, 1521–1550." In *Negotiation within Domination: New Spain's Indian Pueblos Confront the Spanish State,* edited by Ethelia Ruiz Medrano and Susan Kellogg, 19–44. Boulder: University of Colorado Press.

Bacon, Francis. [1625] 1999. *The Essays or Counsels, Civil and Moral.* Edited by Brian Vickers. Oxford: Oxford University Press.

Badiou, Alain. 2007. *Being and Event.* Translated by Oliver Feltham. London: Continuum.

———. 2009. *Logics of Worlds: Being and Event, 2.* London: Continuum.

Bakhtin, M. M. 1968. *Rabelais and His World.* Cambridge, MA: MIT Press.

Balbuena, Bernardo de. [1604] 2011. *Grandeza mexicana.* Edited by Asima F. X. Saad Maura. Madrid: Cátedra.

Banaji, Jairus. 2010. *Theory as History: Essays on Modes of Production and Exploitation.* Boston: Brill.

Barkawi, Tarak. 2010. "State and Armed Force in International Context." In *Mercenaries, Pirates, Bandits and Empires: Private Violence in Historical Context,* edited by Alejandro Colás and Bryan Mabee, 33–53. New York: Columbia University Press.

Barros, João de. [1552] 1998. *Ásia de João de Barros: Dos feitos que os Portugueses fizeram no descobrimento e conquista dos mares e terras do Oriente: Primeira década.* Lisbon: Impresa Nacional–Casa da Moeda.

Bataillon, Marcel. 1966. "La herejía de fray Francisco de la Cruz y la reacción antilascasiana." In *Études sur Bartolomé de Las Casas*, edited by Marcus Raymond, 311–24. Paris: Centre de recherches de l'Institut d'études hispaniques.

Bauer, Ralph. 2003. *The Cultural Geography of Colonial American Literatures: Empire, Travel, Modernity*. Cambridge: Cambridge University Press.

Beaud, Michel. 2001. *A History of Capitalism, 1500–2000*. New York: Monthly Review Press.

Benítez-Rojo, Antonio. 1996. *The Repeating Island: The Caribbean and the Postmodern Perspective*. 2nd ed. Translated by James Maraniss. Durham, NC: Duke University Press.

Benjamin, Walter. 2007. "Theses on the Philosophy of History." In *Illuminations*, edited by Hannah Arendt, 253–64. New York: Schocken Books.

Bentancor, Orlando. 2014. "Francisco de Vitoria, Carl Schmitt, and Originary Technicity." *Política común* 5. http://dx.doi.org/10.3998/pc.12322227.0005.002.

Benton, Lauren. 2009. *A Search for Sovereignty: Law and Geography in European Empires, 1400–1900*. Cambridge: Cambridge University Press.

Beverley, John. 2011. *Latinamericanism after 9/11*. Post-Contemporary Interventions. Durham, NC: Duke University Press.

Blackmore, Josiah. 2002. *Manifest Perdition: Shipwreck Narrative and the Disruption of Empire*. Minneapolis: University of Minnesota Press.

———. 2008. *Moorings: Portuguese Expansion and the Writing of Africa*. Minneapolis: University of Minnesota Press.

Bloch, Ruth H. 1987. "The Gendered Meanings of Virtue in Revolutionary America." *Signs* 13, no. 1: 37–58.

Boccaccio, Giovanni. 1964. *Tutte le Opere di Giovanni Boccaccio*. Vol. 5. Milan: Mondadori.

Bourdieu, Pierre. 1990. *The Logic of Practice*. Translated by Richard Nice. Stanford, CA: Stanford University Press.

———. 1993. *The Field of Cultural Production: Essays on Art and Literature*. Edited by Randal Johnson. New York: Columbia University Press.

———. 2013. *Algerian Sketches*. Cambridge: Polity.

Braudel, Fernand. 1973. *Capitalism and Material Life, 1400–1800*. New York: Harper.

———. 1977. *Afterthoughts on Material Civilization and Capitalism*. Baltimore, MD: Johns Hopkins University Press.

Butler, Judith. 1990. *Gender Trouble: Feminism and the Subversion of Identity*. New York: Routledge.

Campbell, Mary. 1999. *Wonder and Science: Imagining Worlds in Early Modern Europe*. Ithaca, NY: Cornell University Press.

Cañizares-Esguerra, Jorge. 2006a. *Nature, Empire, and Nation: Explorations of the History of Science in the Iberian World*. Stanford, CA: Stanford University Press.

——. 2006b. *Puritan Conquistadors: Iberianizing the Atlantic, 1550–1700*. Stanford, CA: Stanford University Press.

Capéran, Louis. 1912. *Le problème du salut des Infidèles*. Paris: Beauchesne.

Carande Thobar, Ramón. 1965. *Carlos V y sus banqueros*. 3 vols. 2nd ed. Madrid: Sociedad de Estudios y Publicaciones.

Cárdenas Bunsen, José Alejandro. 2011. *Escritura y derecho canónico en la obra de fray Bartolomé de Las Casas*. Madrid: Iberoamericana.

Carrasco, Davíd. 1982. *Quetzalcoatl and the Irony of Empire: Myths and Prophecies in the Aztec Tradition*. Chicago: University of Chicago Press.

Carver, Raymond. 1981. *What We Talk about When We Talk about Love*. New York: Knopf.

Casa de Contratación. 1553. *Ordenanças reales para la Casa de la Contractación de Sevilla, y para otras cosas de las Indias, y de la navegación y contractación dellas*. Seville: Martín de Montesdoca, Repositorio Documental GREDOS. http://hdl.handle.net/10366/49208.

Castiglione, Baldassare. [1528] 2002. *Il Cortigiano*. Edited by Amedeo Quondam. Milan: Mondadori.

Castillero R., Ernesto J. 1957. "Gonzalo Fernández de Oviedo y Valdés, veedor de Tierra Firme." *Revista de Indias* 17, no. 69–70: 521–40.

Castro, Daniel. 2007. *Another Face of Empire: Bartolomé de Las Casas, Indigenous Rights, and Ecclesiastical Imperialism*. Durham, NC: Duke University Press.

Cebrián, Dina Ludeña. 2010. "The Sources and Resources of Our Indigenous Theology." *Ecumenical Review* 62, no. 4: 361–70.

Certeau, Michel de. 1984. *The Practice of Everyday Life*. Translated by Steven Rendall. Berkeley: University of California Press.

——. 1988. *The Writing of History*. Translated by Tom Conley. New York: Columbia University Press.

Cervantes de Salazar, Francisco. [1575] 1914. *Crónica de la Nueva España*. Madrid: Hispanic Society of America.

Cervera Peru, José. 1997. *La Casa de Contratación y el Consejo de Indias: Las razones de un superministerio*. Madrid: Ministerio de Defensa.

Chakrabarty, Dipesh. 2000. *Provincializing Europe: Postcolonial Thought and Historical Difference*. Princeton, NJ: Princeton University Press.

——. 2002. *Habitations of Modernity: Essays in the Wake of Subaltern Studies*. Chicago: University of Chicago Press.

Chaunu, Pierre. 1991. *Conquête et exploitation des Nouveaux Mondes: XVIe siècle*. 4th ed. Paris: Presses Universitaires de France.

Cicero. 1979. *Pro Archia Poeta*. Vol. 11. Translated by N. H. Watts. Cambridge, MA: Harvard University Press.

——. 2001. *Cicero on the Ideal Orator (De Oratore)*. Edited by James M. May. Translated by Jakob Wisse. New York: Oxford University Press.

Cieza de León, Pedro de. [1553] 1994. *Crónica del Peru. Cuarta Parte.* Vol. 3, *Guerra de Quito.* Edited by Laura Gutiérrez Arbulú. Lima: PUCP, Fondo Editorial.

Clarke, Kamari Maxine. 2009. *Fictions of Justice: The International Criminal Court and the Challenges of Legal Pluralism in Sub-Saharan Africa.* Cambridge: Cambridge University Press.

Colás, Alejandro, and Bryan Mabee, eds. 2010. *Mercenaries, Pirates, Bandits, and Empires: Private Violence in Historical Context.* New York: Columbia University Press.

Colón, Cristóbal. 1984. *Textos y documentos completos: Relaciones de viajes, cartas y memoriales.* Edited by Consuelo Varela. Madrid: Alianza.

Colón, Fernando. 1984. *Historia del Almirante.* Edited by Luis Arranz. Translated by Alfonso Ulloa. Madrid: Historia 16.

Columbus, Christopher, and Bartolomé de Las Casas. 1989. *The Diario of Christopher Columbus's First Voyage to America, 1492–1493.* Edited by O. C. Dunn and James E. Kelley. Norman: University of Oklahoma Press.

Conley, Tom. 1992. "De Bry's Las Casas." In *Amerindian Images and the Legacy of Columbus*, edited by René Jara and Nicholas Spadaccini, 103–31. Minneapolis: University of Minnesota Press.

Cornejo Polar, Antonio. 1994. *Escribir en el aire: Ensayo sobre la heterogeneidad socio-cultural en las literaturas andinas.* Lima: Editorial Horizonte.

———. 2013. *Writing in the Air: Heterogeneity and the Persistence of Oral Tradition in Andean Literatures.* Translated by Lynda J. Jentsch. Durham, NC: Duke University Press.

Cortés, Hernán. [1519–26] 1993. *Cartas de relación.* Edited by Angel Delgado Gómez. Madrid: Castalia.

Cotrugli, Benedetto. [1573] 1990. *Il Libro dell'arte di Mercatura.* Edited by Ugo Tucci. Venice: Arsenale.

Covarrubias y Orozco, Sebastián de. 1611. *Tesoro de la lengua castellana o española.* Madrid: Biblioteca Virtual Miguel de Cervantes. www.cervantes virtual.com/.

Curtius, Ernst Robert. 1990. *European Literature and the Latin Middle Ages.* Princeton, NJ: Princeton University Press.

D'Arienzo, Luisa. 2003. *La presenza degli Italiani in Portogallo al tempo di Colombo.* Rome: Istituto Poligrafico e Zecca dello Stato, Libreria dello Stato.

Davenport, Frances Gardiner, and Charles Oscar Paullin, eds. 1917. *European Treaties Bearing on the History of the United States and Its Dependencies.* Washington, DC: Carnegie Institution of Washington.

Davis, Kathleen. 2008. *Periodization and Sovereignty: How Ideas of Feudalism and Secularization Govern the Politics of Time.* Philadelphia: University of Pennsylvania Press.

Day, Gerald W. 1988. *Genoa's Response to Byzantium, 1155–1204: Commercial Expansion and Factionalism in a Medieval City.* Urbana: University of Illinois Press.

DeGuzmán, María. 2005. *Spain's Long Shadow: The Black Legend, Off-Whiteness, and Anglo-American Empire.* Minneapolis: University of Minnesota Press.

Deive, Carlos Esteban. 1995. *La Española y la esclavitud del indio.* Santo Domingo: Fundación García Arévalo.

Deleuze, Gilles, and Félix Guattari. 1987. *A Thousand Plateaus.* Translated by Brian Massumi. Minneapolis: University of Minnesota Press.

Del Valle, Ivonne. 2009. *Escribiendo desde los márgenes: Colonialismo y jesuitas en el siglo XVIII.* Mexico City: Siglo XXI.

———. 2012. "José de Acosta, Violence and Rhetoric: The Emergence of the Colonial Baroque." *Calíope: Journal of the Society for Renaissance and Baroque Hispanic Society* 18, no. 2: 46–47.

De Man, Paul. 1979. *Allegories of Reading: Figural Language in Rousseau, Nietzsche, Rilke, and Proust.* New Haven, CT: Yale University Press.

Denglos, J. 1992. Introduction to *Doce dudas* by Bartolomé de Las Casas, edited by J. B. Lassegue, v–xxxiii. Madrid: Alianza.

Derrida, Jacques. 1994. *Specters of Marx: The State of the Debt, the Work of Mourning, and the New International.* New York: Routledge.

———. 1997. *Of Grammatology.* Translated by Gayatri Chakravorty Spivak. Baltimore, MD: Johns Hopkins University Press.

———. 2005. *The Politics of Friendship.* 2nd ed. Translated by George Collins. New York: Verso.

———. 2009. *The Beast and the Sovereign.* Edited by Michel Lisse, Marie-Louise Mallet, and Ginette Michaud. Translated by Geoffrey Bennington. Chicago: University of Chicago Press.

Detienne, Marcel. 2003. *Comment être autochtone: Du pur Athénien au François raciné.* Paris: Seuil.

Díaz del Castillo, Bernal. [1568/1632] 1984. *Historia verdadera de la conquista de la Nueva España.* 2 vols. Edited by Miguel León-Portilla. Madrid: Historia 16.

Diccionario de Autoridades. 1724. s.v. "mancomunidad," "conquista," "empresa." Nuevo Tesoro Lexicográfico by Real Academia Española. http://ntlle .rae.es/.

Drelichman, Mauricio, and Hans-Joachim Voth. 2014. *Lending to the Borrower from Hell: Debt, Taxes, and Default in the Age of Philip II.* Princeton, NJ: Princeton University Press.

Dussel, Enrique. 2007. "Alterity and Modernity (Las Casas, Vitoria, and Suárez: 1514–1617)." Translated by James Terry. In *Postcolonialism and Political Theory*, edited by Nalini Persram, 3–36. Lanham, MD: Lexington Books.

Duval, Edwin M. 1991. *The Design of Rabelais's "Pantagruel."* New Haven, CT: Yale University Press.

Duviols, Pierre. 1977. *La destrucción de las religiones andinas (durante la conquista y la colonia)*. Translated by Albor Maruenda. Mexico City: UNAM.

Elias, Norbert. 2000. *The Civilizing Process: Sociogenetic and Psychogenetic Investigations*. Edited by Johan Goudsblom and Stephen Mennell. Translated by Eric Dunning. Oxford: Blackwell.

Ercilla, Alonso de. [1569–89/90] 1993. *La Araucana*. Edited by Isaías Lerner. Madrid: Cátedra.

Esposito, Roberto. 2013. *Terms of the Political: Community, Immunity, Biopolitics*. Edited by Rhiannon Noel Welch. Commonalities. New York: Fordham University Press.

Estella, Diego de. 1775. *Tratado de la vanidad del mundo y al fin un tratado del Amor de Dios*. Madrid: Don Pedro Marin.

Federici, Silvia. 2004. *Caliban and the Witch: Women, the Body, and Primitive Accumulation*. Brooklyn, NY: Autonomedia.

Fernández de Enciso, Martín. 1578. *A Briefe Description of the Portes, Creekes, Bayes, and Hauens, of the Weast India: Translated out of the Castlin Tongue by I.F. The Originall Whereof Was Directed to the Mightie Prince Don Charles, King of Castile, &c.* London: Henry Bynneman. http://gateway.proquest.com.

———. [1519] 1987. *Suma de geografía*. Edited by Mariano Cuesta Domingo. Madrid: Museo Naval.

Fernández de Oviedo y Valdés, Gonzalo. [1526] 1950. *Sumario de la natural historia de las Indias*. Mexico City: Fondo de Cultura Económica.

———. [1535] 1959. *Historia general y natural de las Indias*. 5 vols. Edited by José de Tudela Bueso. Madrid: Atlas.

———. [1878] 1989. *Batallas y quinquagenas*. Edited by Juan Bautista Avalle-Arce. Madrid: Ediciones de la Diputación de Salamanca.

———. [1519] 2002. *Claribalte*. Edited by María José Rodilla. Mexico City: UNAM.

Fisher, John Robert. 1977. *Silver Mines and Silver Miners in Colonial Peru, 1776–1824*. Liverpool: Centre for Latin American Studies, University of Liverpool.

Fita, Fidel, ed. 1887. "Edicto de los Reyes Católicos (31 Marzo, 1492): Desterrando de sus estados a todos los judíos." *Boletín de la Real Academia de la Historia* 11, no. 1–3 (July–September): 512–28.

Foucault, Michel. 1997. "Security, Territory and Population." In *The Essential Works of Foucault, 1954–1984*, edited by Paul Rabinow, 1:67–71. New York: New Press.

Freeman, John. 2005. "Venture Capitalism and Modern Capitalism." In *The Economic Sociology of Capitalism*, edited by Victor Nee and Richard Swedberg, 144–67. Princeton, NJ: Princeton University Press.

Freud, Sigmund. 1997. *Writings on Art and Literature*. Stanford, CA: Stanford University Press.

Frye, Northrop. 2006. *Anatomy of Criticism: Four Essays*. Edited by Robert D. Denham. Toronto: University of Toronto Press.

Fuchs, Barbara. 2001. *Mimesis and Empire: The New World, Islam, and European Identities*. Cambridge: Cambridge University Press.

Galindo, Alberto Flores. 1986. *Europa y el país de los incas: La utopía andina*. Lima: Instituto de Apoyo Agrario.

Garcilaso de la Vega, el Inca. [1609] 1998. *Comentarios reales, primera parte*. Mexico City: Editorial Porrúa.

Garin, Eugenio. 1969. *Science and Civic Life in the Italian Renaissance*. Garden City, NY: Anchor Books.

Gaylord, Mary M. 1999. "The True History of Early Modern Writing in Spanish: Some American Reflections." In *The Places of History: Regionalism Revisited in Latin America*, edited by Doris Sommer, 81–93. Durham, NC: Duke University Press.

Genette, Gérard. 2004. *Métalepse: De la figure à la fiction*. Poétique. Paris: Seuil.

Gerbi, Antonello. 1949. "El *Claribalte* de Oviedo." *Fénix* 5–6: 378–90.

Gibson, Charles. 1977. "Reconquista and Conquista." In *Homage to Irving A. Leonard: Essays on Hispanic Art, History and Literature*, edited by Donald A. Yates and Raquel Chang-Rodríguez, 19–28. Ann Arbor: Latin American Studies Center, Michigan State University Press.

Goldwert, Marvin. 1955. "La lucha por la perpetuidad de las encomiendas en el Peru Virreinal (1550–1600)." *Revista Histórica* 22: 336–60.

———. 1957. "La lucha por la perpetuidad de las encomiendas en el Perú Virreinal, 1550–1600 (Continuación)." *Revista Histórica* 23: 207–45.

González Echevarría, Roberto. 1998. *Myth and Archive: A Theory of Latin American Narrative*. 2nd ed. Durham, NC: Duke University Press.

González Ruiz, José María. 1962. "Extra Ecclesiam Nulla Salus a la luz de la teología paulina." *Selecciones de Teología*, no. 4: 282–84.

Gose, Peter. 2008. *Invaders as Ancestors: On the Intercultural Making and Unmaking of Spanish Colonialism in the Andes*. Toronto: University of Toronto Press.

Graeber, David. 2014. *Debt: The First Five Thousand Years*. New York: Penguin Random House.

Grafe, Regina, and Alejandra Irigoin. 2012. "A Stakeholder Empire: The Political Economy of Spanish Imperial Rule in America." *Economic History Review* 65, no. 2: 609–51.

Gramsci, Antonio. 1997. *Selections from the Prison Notebooks*. Edited and translated by Quentin Hoare and Geoffrey Nowell Smith. New York: International Publishers.

Green, Keith. 1992. "Deixis and the Poetic Persona." *Language and Literature* 1, no. 2: 121–34.

Greenblatt, Stephen. 1991. *Marvelous Possessions: The Wonder of the New World.* Chicago: University of Chicago Press.

———. [1980] 2005. *Renaissance Self-Fashioning: From More to Shakespeare.* 2nd ed. Chicago: University of Chicago Press.

Greene, Roland Arthur. 1999. *Unrequited Conquests: Love and Empire in the Colonial Americas.* Chicago: University of Chicago Press.

Greer, Margaret Rich, Walter Mignolo, and Maureen Quilligan, eds. 2007. *Rereading the Black Legend: The Discourses of Religious and Racial Difference in the Renaissance Empires.* Chicago: University of Chicago Press.

Gruzinski, Serge. 2002. *The Mestizo Mind: The Intellectual Dynamics of Colonization and Globalization.* New York: Routledge.

———. 2014. *The Eagle and the Dragon: European Expansion and Globalization in the 16th Century.* Malden, MA: Polity.

Guamán Poma de Ayala, Felipe. [1615/16] 2001. *El primer nueva corónica y buen gobierno (1615/1616).* Autograph manuscript facsimile, annotated transcription, documents, and other digital resources. The Guaman Poma Website: A Digital Research Center of the Royal Library, Copenhagen, Denmark. www.kb.dk/permalink/2006/poma/info/en/frontpage.htm.

Guha, Ranajit. 1997. *Dominance without Hegemony: History and Power in Colonial India.* Cambridge, MA: Harvard University Press.

———. [1983] 1999. *Elementary Aspects of Peasant Insurgency in Colonial India.* Durham, NC: Duke University Press.

Gutiérrez, Gustavo. 1993. *Las Casas: In Search of the Poor of Jesus Christ.* Translated by Robert R. Barr. Maryknoll, NY: Orbis Books.

Hackett, Charles W., ed. 1923. *Historical Documents Relating to New Mexico, Nueva Vizcaya, and Approaches Thereto, to 1773.* Vol. 1. Washington, DC: Carnegie Institute of Washington.

Hanke, Lewis. 1935. *The First Social Experiments in America: A Study in the Development of Spanish Indian Policy in the Sixteenth Century.* Cambridge, MA: Harvard University Press.

———. 1938. "The 'Requerimiento' and Its Interpreters." *Revista de Historia de América* 1: 25–34.

———. 1949. *The Spanish Struggle for Justice in the Conquest of America.* Dallas, TX: Southern Methodist University Press.

———. 1974. *All Mankind Is One: A Study of the Disputation between Bartolomé de Las Casas and Juan Ginés de Sepúlveda in 1550 on the Intellectual and Religious Capacity of the American Indians.* DeKalb: Northern Illinois University Press.

Hardt, Michael, and Antonio Negri. 2000. *Empire.* Cambridge, MA: Harvard University Press.

Hariot, Thomas. 1590. *A Briefe and True Report of the New Found Land of Virginia.* Available at Adam Matthew, Marlborough, Empire Online, www.empire.amdigital.co.uk.

Harvey, David. 2001. *Spaces of Capital: Towards a Critical Geography*. Edinburgh: Edinburgh University Press.

Hobbes, Thomas. [1651] 2012. *Leviathan*. Oxford: Clarendon Press.

Horswell, Michael J. 2005. *Decolonizing the Sodomite: Queer Tropes of Sexuality in Colonial Andean Culture*. Austin: University of Texas Press.

Huarte de San Juan, Juan. [1575] 1989. *Examen de ingenios para las ciencias*. Edited by Guillermo Serés. Madrid: Cátedra.

Huntington, Samuel P. 2011. *The Clash of Civilizations and the Remaking of World Order*. New York: Simon & Schuster.

Jáuregui, Carlos. 2008. *Canibalia: Canibalismo, calibanismo, antropofagia cultural y consumo en América Latina*. Madrid: Iberoamericana.

———. 2009. "Cannibalism, the Eucharist, and Criollo Subjects." In *Creole Subjects in the Colonial Americas: Empires, Texts, Identities*, edited by Ralph Bauer and Jose Antonio Mazzotti, 61–100. Chapel Hill: University of North Carolina Press.

John of the Cross. 2009. *San Juan de la Cruz*. Edited by Francisco Javier Díez de Revenga. Madrid: Fundación José Antonio de Castro.

Johnson, Christine. 2008. *The German Discovery of the World: Renaissance Encounters with the Marvelous*. Charlottesville: University of Virginia Press.

Johnson, Walter. 2013. *River of Dark Dreams: Slavery and Empire in the Cotton Kingdom*. Cambridge, MA: Harvard University Press.

Kamen, Henry. 2004. *Empire: How Spain Became a World Power, 1492–1763*. New York: Perennial.

Kant, Immanuel. 2015. *On the Perpetual Peace: A Philosophical Sketch*. Edited by Brian Orend. Translated by Ian Johnston. Peterborough, Ont.: Broadview Press.

Kellebenz, Hermann. 1978. "Los Fugger en España en la época de Felipe II: ¿Fue un buen negocio el arrendamiento de los maestrazgos después de 1562?" In *Dinero y crédito (siglos XVI y XIX): Actas del Primer Coloquio Internacional de Historia Económica*, edited by Alfonso Otazu, 19–36. Madrid: Artes Gráficas Benzal.

Kellogg, Susan. 2010. "Back to the Future: Law, Politics, and Culture in Colonial Mexican Ethnohistorical Studies." In *Negotiation within Domination: New Spain's Indian Pueblos Confront the Spanish State*, edited by Ethelia Ruíz Medrano and Susan Kellogg, 1–18. Boulder: University of Colorado Press.

Konetzke, Richard, ed. 1953. *Colección de documentos para la historia de la formación social de Hispanoamérica, 1493–1810*. Vol. 1. Madrid: Consejo Superior de Investigaciones Científicas.

Kuriyama, Shigehisa. 2002. "The Enigma of 'Time Is Money.'" *Japan Review* 14: 217–30.

Lafaye, Jacques. 1976. *Quetzalcóatl and Guadalupe: The Formation of Mexican National Consciousness, 1531–1813*. Chicago: University of Chicago Press.

Lanning, Edward P. 1967. *Peru before the Incas*. Englewood Cliffs, NJ: Prentice-Hall, Spectrum Book.

Las Casas, Bartolomé de. [1527–61] 1967. *Apologética historia sumaria*. Edited by Edmundo O'Gorman. 2 vols. Mexico City: UNAM.

———. [1518–66] 1988–. *Obras completas*. 14 vols. Edited by Alvaro Huerga. Madrid: Alianza.

———. [ca. 1538] 1990. *De unico vocationis modo*. Edited by Paulino Castañeda Delgado. *Obras completas*, vol. 2. Madrid: Alianza.

———. [1552] 1992a. *Tratados de 1552 impresos por Las Casas en Sevilla*. Edited by Ramón Hernández Martín and Lorenzo Galmés Mas. *Obras completas*, vol. 10. Madrid: Alianza.

———. [ca. 1563] 1992b. *De Thesauris*. Edited by Ángel Losada. *Obras completas*, vol. 11.1. Madrid: Alianza.

———. [ca. 1564] 1992c. *Doce dudas*. Edited by J. B. Lassegue. *Obras completas*, vol. 11.2. Madrid: Alianza.

———. [1571] 1992d. *De regia potestate*. Edited by Jaime González Rodríguez. *Obras completas*, vol. 12. Madrid: Alianza.

———. [1527–61] 1994. *Historia de las Indias*. Edited by Miguel Angel Medina, Jesús Angel Barreda, and Isacio Pérez Fernández. *Obras completas*, vols. 3–5. Madrid: Alianza.

———. [1516–66] 1995. *Cartas y memoriales*. Edited by Paulino Castañeda, Carlos de Rueda, Carmen Godínez, and Inmaculada de La Corte. *Obras completas*, vol. 13. Madrid: Alianza.

———. [1552] 2005. *Brevísima relación de la destruición de las Indias*. Edited by André Saint-Lu. Madrid: Cátedra.

Las Casas, Bartolomé de, and Domingo de Santo Tomás. [1560] 1957. *Memorial del obispo fray Bartolomé de Las Casas y fray Domingo de Santo Tomás (1560)*. In *Obras escogidas*, edited by José Tudela y Bueso, 465–68. Madrid: Biblioteca de Autores Españoles.

Las Casas Grive, Mercedes de, ed. [1553] 2003. *Relación de las cosas acaescidas en las alteraciones del Perú después que Blasco Núñez entró en él*. Lima: Editorial Pontificia Universidad Católica del Perú.

Latour, Bruno. 1999. *Pandora's Hope: Essays on the Reality of Science Studies*. Cambridge, MA: Harvard University Press.

Lea, Henry Charles. 2011. *A History of the Inquisition of Spain: And the Inquisition in the Spanish Dependencies*. Edited by Lu Ann Homza. London: I. B. Tauris.

Le Goff, Jacques. 1988. *Your Money or Your Life: Economy and Religion in the Middle Ages*. New York: Zone Books.

León, Luis de. [1583] 1977. *De los nombres de Cristo*. Madrid: Cátedra.

Leonard, Irving. 1992. *Books of the Brave: Being an Account of Books and Men in the Spanish Conquest and Settlement of the Sixteenth-Century New World*. Berkeley: University of California Press.

León-Portilla, Miguel. 2001. *La huida de Quetzalcóatl.* Mexico City: Fondo de Cultura Económica, 2001.

Levillier, Roberto, ed. 1921. *Gobernantes del Perú, cartas y papeles, siglo XVI.* Madrid: Sucesores de Rivadeneyra.

Lévi-Strauss, Claude. 1962. *La pensée sauvage.* Paris: Plon.

———. 1997. *Tristes tropiques.* New York: Modern Library.

Lisi, Francesco Leonardo, and the Catholic Church, Province of Peru, Concilio Provincial (3rd: 1582–1583). 1990. *El Tercer Concilio Limense y la aculturación de los indígenas sudamericanos estudio crítico con edición, traducción y comentario de las actas del concilio provincial celebrado en Lima entre 1582 y 1583.* Salamanca: Universidad de Salamanca.

Lockhart, James. 1972. *The Men of Cajamarca: A Social and Biographical Study of the First Conquerors of Peru.* Austin: University of Texas Press.

———. 1994. *Spanish Peru, 1532–1560: A Social History.* Madison: University of Wisconsin Press.

Lohmann Villena, Guillermo. 1977. *Las ideas jurídico-políticas en la rebelión de Gonzalo Pizarro: La tramoya doctrinal del levantamiento contra las Leyes Nuevas en el Perú.* Valladolid: Casa-Museo de Colón y Seminario Americanista.

———. 1999. *Las minas de Huancavelica en los siglos XVI y XVII.* Lima: Editorial Pontificia Universidad Católica del Perú.

Lopez, Robert Sabatino, and Irving Woodworth Raymond, eds. 2001. *Medieval Trade in the Mediterranean World: Illustrative Documents.* Translated by Olivia Remie Constable. New York: Columbia University Press.

López de Gómara, Francisco. [1553] 1987. *La conquista de México.* Edited by José Luis de Rojas. Madrid: Historia 16.

López Hernández, Eleazar. 2000. *Teología India: Antología.* Cochabamba: Verbo Divino.

Lordon, Frederic. 2014. *Willing Slaves of Capital: Spinoza and Marx on Desire.* New York: Verso.

Losada, Ángel. 1992. Introduction to *De Thesauris* by Bartolomé de Las Casas, edited by Ángel Losada and Martín Lassègue, i–xxviii. Madrid: Alianza.

MacCormack, Sabine. 1988. "Pachacuti: Miracles, Punishments, and Last Judgment: Visionary Past and Prophetic Future in Early Colonial Peru." *American History Review* 93, no. 4: 960–1006.

———. 1991. *Religion in the Andes: Vision and Imagination in Early Colonial Peru.* Princeton, NJ: Princeton University Press.

Machiavelli, Niccolò. [1532] 1995. *Il Principe.* Edited by Giorgio Inglese. Turin: Einaudi.

———. [1531] 2000. *Discorsi sopra la Prima Decada di Tito Livio.* Edited by Corrado Vivanti. Turin: Einaudi.

Malinowski, Bronislaw. [1925] 1992. *Magic, Science, and Religion, and Other Essays*. Prospect Heights, IL: Waveland Press.

Manrique, Nelson. 2002. *El tiempo del miedo: La violencia política en el Perú, 1980–1996*. Lima: Fondo Editorial del Congreso del Perú.

Maravall, José Antonio. 1999. *Carlos V y el pensamiento político del renacimiento*. Madrid: Boletín Oficial del Estado, Centro de Estudios Políticos y Constitucionales.

Martínez, María Elena. 2011. *Genealogical Fictions: Limpieza de Sangre, Religion, and Gender in Colonial Mexico*. Stanford, CA: Stanford University Press.

Martínez-San Miguel, Yolanda. 2017. "Colonial and Mexican Archipelagoes: Reimagining Colonial Caribbean Studies." In *Archipelagic American Studies*, edited by Brian Russell Roberts and Michelle Ann Stephens, 155–74. Durham, NC: Duke University Press.

Maticorena-Estrada, M. 1966. "El contrato de Panamá, 1526, para el Descubrimiento del Perú." *Cahiers du monde hispanique et luso-brasilien* 7: 55–84.

Marx, Karl. [1867] 1992–93. *Capital: A Critique of Political Economy*. 3 vols. Translated by Ben Fowkes and David Fernbach. London: Penguin and New Left Review.

Marx, Karl, and Friedrich Engels. [1852] 1988. *The Eighteenth Brumaire of Louis Bonaparte, with Explanatory Notes*. Translated by Daniel De Leon. New York: International Publishers.

Mauss, Marcel. [1902] 2001. *A General Theory of Magic*. London: Routledge.

Mbembe, Achille. 2003. "Necropolitics." *Public Culture* 15, no. 1: 11–40.

McBrien, Justin. 2016. "Accumulating Extinction: Planetary Catastrophism in the Necrocene." In *Anthropocene or Capitalocene? Nature, History, and the Crisis of Capitalism*, edited by Jason W. Moore, 116–37. Oakland, CA: PM Press.

McClintock, Anne. 1995. *Imperial Leather: Race, Gender, and Sexuality in the Colonial Conquest*. New York: Routledge.

Melville, Stephen, and Bill Readings, eds. 1995. *Vision and Textuality*. Durham, NC: Duke University Press.

Menocal, María Rosa. 1994. *Shards of Love: Exile and the Origins of the Lyric*. Durham: Duke University Press.

Merrim, Stephanie. 1982. "The Castle of Discourse: Fernández de Oviedo's *Don Claribalte* (1519) or 'Los correos andan más que los caballeros.'" *Modern Language Notes* 97: 329–46.

Mignolo, Walter. 1981. "El metatexto historiográfico y la historiografía indiana." *Modern Language Notes* 90, no. 2: 358–402.

———. 2002. Introduction to *Natural and Moral History of the Indies*, by José de Acosta, xvii–xxviii. Edited by Jane E. Mangan. Translated by Frances M. López-Morillas. Durham, NC: Duke University Press.

Miller, Jacques-Alain. 1977–78. "Suture (Elements of the Logic of the Signifier)." Translated by Jacqueline Rose. *Screen* 18, no. 4: 24–34.

Mira Caballos, Esteban. 1997. *El indio antillano: Repartimiento, encomienda y esclavitud (1492–1542).* Seville: Múnoz Moya Editores.

Molina, Cristóbal de. [1573] 1989. *Fábulas y ritos de los Incas.* Edited by Henrique Urbano and Pierre Duviols. Madrid: Historia 16.

Montenegro, Giovanna. 2017. "Conquistadors and Indians 'Fail' at Gift Exchange: An Analysis of Nikolaus Federmann's Indianische Historia (Haguenau, 1557)." *Modern Language Notes* 132, no. 2: 272–90.

Moore, Jason W. 2009. "Madeira, Sugar, and the Conquest of Nature in the First Sixteenth Century, Part I." *Review* 32, no. 4: 345–90.

———. 2010. "Madeira, Sugar, and the Conquest of Nature in the First Sixteenth Century, Part II." *Review* 33, no. 1: 1–24.

———. 2015. *Capitalism in the Web of Life.* London: Verso.

———. 2016. "The Rise of Cheap Nature." In *Anthropocene or Capitalocene? Nature, History and the Rise of Capitalism,* edited by Jason W. Moore, 78–116. Oakland, CA: PM Press.

Morales Padrón, Francisco. 2008. *Teoría y leyes de la Conquista.* Seville: Universidad de Sevilla.

Moraña, Mabel, Enrique D. Dussel, and Carlos A. Jáuregui, eds. 2008. *Coloniality at Large: Latin America and the Postcolonial Debate.* Durham, NC: Duke University Press.

More, Anna. 2014. "From Lines to Networks: Carl Schmitt's *Nomos* and the Early Atlantic System." *Política Común* 5. http://dx.doi.org/10.3998/pc.12322227.0005.004.

Moreiras, Alberto. 2000. "Ten Notes on Primitive Imperial Accumulation: Ginés de Sepúlveda, Las Casas, Fernández de Oviedo." *Interventions* 2, no. 3: 343–63.

Morison, Samuel Eliot. 1942. *Admiral of the Ocean Sea: A Life of Christopher Columbus.* Boston, MA: Northeastern University Press.

Moscoso, Francisco. 1986. *Tribu y clases en el caribe antiguo.* San Pedro de Macorís: Universidad Central del Este.

Muldoon, James. 1980. "John Wyclif and the Rights of the Infidels: The *Requerimiento* Reexamined." *The Americas* 36, no. 3: 301–16.

Murra, John V. 2002. *El mundo andino: Población, medio ambiente y economía.* Lima: Editorial Pontificia Universidad Católica del Perú.

Murúa, Martín de. [1616] 1987. *Historia general del Perú.* Edited by Manuel Ballesteros Gaibrois. Madrid: Historia 16.

———. 2008. *Historia general del Piru: Facsimile of J. Paul Getty Museum Ms. Ludwig XIII 16.* Los Angeles: Getty Research Institute.

Navarrete, Ignacio Enrique. 1994. *Orphans of Petrarch: Poetry and Theory in the Spanish Renaissance.* Berkeley: University of California Press.

Nebrija, Antonio de. [1492] 1931. *Gramática castellana por D. Antonio de Nebrija.* Edited by José Rogerio Sánchez. Madrid: Editorial Hernando.

Nederman, Cary J. 2000. "Machiavelli and Moral Character: Principality, Republic and the Psychology of *Virtù*." *History of Political Thought* 21, no. 3: 349–64.

Negri, Antonio. 2008. *The Porcelain Workshop: For a New Grammar of Politics*. Translated by Noura Wedell. Los Angeles: Semiotext(e).

Nemser, Daniel. 2014. "Primitive Spiritual Accumulation and the Extraction Economy." *Política Común* 4. http://dx.doi.org/10.3998/pc.12322227.0005.003.

——. 2017. *Infrastructures of Race: Concentration and Biopolitics in Colonial Mexico*. Austin: University of Texas Press.

Nerlich, Michael. 1987. *Ideology of Adventure: Studies in Modern Consciousness, 1100–1750*. Minneapolis: University of Minnesota Press.

Nicholas of Cusa. 1989. *The Layman on Wisdom and the Mind*. Translated by M. L. Führer. Ottawa: Dovehouse.

Noonan, John, Jr. 1957. *The Scholastic Analysis of Usury*. Cambridge, MA: Harvard University Press.

Nunes Dias, Manuel. 1963. *O capitalismo monárquico português (1415–1549): Contribuição o estudo das origens do capitalismo moderno*. 2 vols. Coimbra: Faculdade de Letras da Universidade de Coimbra.

Núñez Cabeza de Vaca, Alvar. [1555] 1998. *Naufragios*. Edited by Juan Francisco Maura. Madrid: Cátedra.

——. [1542] 1999. *Alvar Núñez Cabeza de Vaca: His Account, His Life, and the Expedition of Pánfilo de Narváez*. Edited by Rolena Adorno and Patrick Charles Pautz. Lincoln: University of Nebraska Press.

——. [1542] 2003. *The Narrative of Cabeza de Vaca*. Edited and translated by Rolena Adorno and Patrick Charles Pautz. Lincoln: University of Nebraska Press.

Obeyesekere, Gananath. 1997. *The Apotheosis of Captain Cook: European Mythmaking in the Pacific*. Princeton, NJ: Princeton University Press and Bishop Museum Press.

O'Gorman, Edmundo. 1967. Introduction to *Apologética historia sumaria*, by Bartolomé de Las Casas, 1:vii–xli. 2 vols. Edited by Edmundo O'Gorman. Mexico City: UNAM.

——. [1958] 2006. *La invención de América: Investigación acerca de la estructura histórica del Nuevo Mundo y del sentido de su devenir*. Mexico City: Fondo de Cultura Económica.

Ollman, Bertell. 1976. *Alienation*. New York: Cambridge University Press.

Otte, Enrique. 1978. "Sevilla, Plaza Bancaria Europea en el siglo XVI." In *Dinero y crédito (siglos XVI y XIX): Actas del Primer Coloquio Internacional de Historia Económica*, edited by Alfonso Otazu, 89–114. Madrid: Artes Gráficas Benzal.

——. 2008. *Sevilla, Siglo XVI: Materiales para su historia económica*. Seville: Centro de Estudios Andaluces.

Owensby, Brian. 2010. Foreword to *Negotiation within Domination: New Spain's Indian Pueblos Confront the Spanish State*, edited by Ethelia Ruíz Medrano and Susan Kellogg, xi–xv. Boulder: University of Colorado Press.

Pagden, Anthony. 1982. *The Fall of Natural Man: The American Indian and the Origins of Comparative Ethnology*. Cambridge: Cambridge University Press.

Parenti, Christian. 2016. "Environment-Making in the Capitalocene: Political Ecology of the State." In *Anthropocene or Capitalocene? Nature, History, and the Crisis of Capitalism*, edited by Jason W. Moore, 166–84. Oakland, CA: PM Press.

Parish, Helen Rand. 1967. *The Life and Writings of Bartolomé de Las Casas*. Albuquerque: University of New Mexico Press.

Parish, Helen Rand, and Harold E. Weidman. 1992. *Las Casas en México: Historia y obras desconocidas*. Mexico City: Fondo de Cultura Económica.

Pastor, Beatriz. 1992. *The Armature of Conquest: Spanish Accounts of the Discovery of America, 1492–1589*. Stanford, CA: Stanford University Press.

Patterson, Orlando. 1982. *Slavery and Social Death: A Comparative Study*. Cambridge, MA: Harvard University Press.

Paz, Octavio. 1990. *Sor Juana, or, The Traps of the Faith*. Translated by Margaret Sayers Peden. Cambridge, MA: Harvard University Press.

Pérez de Oliva, Hernán. [ca. 1583] 1965. *Historia de la invención de las Yndias*. Edited by José Juan Arrom. Bogotá: Instituto Caro y Cuervo.

Pérez de Tudela Bueso, Juan. 1957. "Rasgos del semblante espiritual de Gonzalo Fernández de Oviedo: La hidalguía caballeresca ante el Nuevo Mundo." *Revista de Indias* 17, no. 69–70: 391–443.

Pérez Luño. 1990. Introduction to *De regia potestate*, edited by Jaime González Rodríguez, i–lix. Madrid: Alianza.

Perlin, John. 1989. *A Forest Journey: The Role of Wood in the Development of Civilization*. New York: Norton.

Perroux, François. 1948. *Le capitalisme*. Paris: Presses Universitaires de France.

Petrarch. 1976. *Petrarch's Lyric Poems: The Rime Sparse and Other Lyrics*. Edited and translated by Robert M. Durling. Cambridge, MA: Harvard University Press.

———. [1346] 1992. *De vita solitaria*. Edited by Marco Noce. Milan: Mondadori.

Pignarre, Philippe, and Isabelle Stengers. 2005. *La sorcellerie capitaliste: Pratique de désenvoûtement*. Paris: La Découverte.

Pinet, Simone. 2010. "Where One Stands: Shipwreck, Perspective, and Chivalric Fiction." *eHumanista: Journal of Medieval and Early Modern Iberian Studies* 16: 381–94.

Plato. 1977. *The Portable Plato: Protagoras, Symposium, Phaedo, and the Republic, Complete, in the English Translation of Benjamin Jowett*. New York: Penguin.

Poleggi, Ennio. 2008. *Ritratto di Genova nel '400: Veduta d'invenzione.* Genoa: Sagep.

Powell, Philip Wayne. 2008. *Tree of Hate: Propaganda and Prejudices Affecting United States Relations with the Hispanic World.* Albuquerque: University of New Mexico Press.

Price, Russell. 1970. "Virtu in Machiavelli's *Il Principe* and *Discorsi.*" *Political Science* 22, no. 2: 43–49.

Pugh, Jonathan. 2013. "Island Movements: Thinking with the Archipelago." *Island Studies Journal* 8, no. 1: 9–24.

Quintilian. 1970. *The Orator's Education.* Vols. 124–27. Loeb Classical Library. Cambridge, MA: Harvard University Press.

Rabasa, José. 1993. *Inventing America: Spanish Historiography and the Formation of Eurocentrism.* Norman: University of Oklahoma Press.

———. 2000. *Writing Violence on the Northern Frontier: The Historiography of Sixteenth-Century New Mexico and Florida and the Legacy of Conquest.* Durham, NC: Duke University Press.

———. 2010. *Without History: Subaltern Studies, the Zapatista Insurgency, and the Specter of History.* Pittsburgh, PA: University of Pittsburgh Press.

———. 2011. *Tell Me the Story of How I Conquered You: Elsewheres and Ethnosuicide in the Mesoamerican World.* Austin: University of Texas Press.

Rabelais, François. [1534] 1996. *Gargantua.* Edited by Guy Demerson, Michel Renaud, and Geneviève Demerson. Paris: Seuil.

———. [ca. 1532] 1996. *Pantagruel.* Edited by Guy Demerson, Michel Renaud, and Geneviève Demerson. Paris: Seuil.

Rama, Angel. [1984] 1996. *The Lettered City.* Edited and translated by John Charles Chasteen. Durham, NC: Duke University Press.

Rappaport, Joanne, and Tom Cummins. 2012. *Beyond the Lettered City: Indigenous Literacies in the Andes.* Durham, NC: Duke University Press.

Ratzinger, Joseph. 1963. *Salus Extra Ecclesiam Nulla Est.* Rome: IDOC, 1963.

Rawls, John. 1999. *The Law of Peoples: with, "The Idea of Public Reason Revisited."* Cambridge, MA: Harvard University Press.

R. B. 1777. *The english hero: or, Sir Francis Drake, revived. Being a full account of the dangerous voyages, admirable adventures, notable discoveries, and magnanimous Atchievements of that valiant and renowned Commander. I. His Voyage in 1572, to Nombre de Dios, in the West-Indies, where they saw a Pile of Silver, Bars near seventy Feet long, ten Feet broad, and twelve Feet high. II. His encompassing the whole World in 1577, which he performed in two Years and ten Months, gaining a vast quantity of Gold and Silver. III. His Voyage into America in 1585, and taking the Towns of St. Jago, St. Domingo, Carthagena, and St. Augustine; also his worthy Actions when Vice-Admiral of England in the Spanish Invasion, 1588. IV. His last voyage into those Countries in 1595, with the Manner of his Death and Burial. Recommended to the Imitation of all heroic Spirits. Enlarged and reduced into chapters with contents.* Eighteenth Century Collections Online. Gale. http: //find.gale.com.ezproxy.princeton.edu/ecco/infomark.do?&source=gale &prodId=ECCO&userGroupName=prin77918&tabID=T001&docId =CW101123962&type=multipage&contentSet=ECCOArticles&version =1.0&docLevel=FASCIMILE>.

Reséndez, Andrés. 2016. *The Other Slavery: The Uncovered Story of Indian Enslavement in America.* New York: Houghton-Mifflin Harcourt.

Roberts, Brian Russell, and Michelle Ann Stephens, eds. 2017. *Archipelagic American Studies.* Durham, NC: Duke University Press.

Roilos, Panagiotis, and Dimitrios Yatromanolakis. 2003. *Ritual Poetics in Greek Culture.* Cambridge, MA: Harvard University Press.

Rougement, Denis de. 1974. *Love in the Western World.* Translated by Montgomery Belgion. New York: Harper & Row.

Rousseau, Jean-Jacques. [1762] 1978. *On the Social Contract with Geneva Manuscript and Political Economy.* Translated by Judith R. Masters. New York: St. Martin's.

Ruhstaller, Stefan. 1992. "Bartolomé de las Casas y su copia del *Diario de a Bordo* de Colón." *Cauce,* no. 14–15: 615–37.

Ruíz Medrano, Ethelia. 2010. "Fighting Destiny: Nahua Nobles and Friars in the Sixteenth-Century Revolt of the Encomenderos against the King." In *Negotiation within Domination: New Spain's Indian Pueblos Confront the Spanish State,* edited by Ethelia Ruíz Medrano and Susan Kellogg, 45–78. Boulder: University of Colorado Press.

Saco, José Antonio. 1932. *Historia de la esclavitud de los indios en el nuevo mundo seguida de la Historia de los repartimientos y encomiendas.* 2 vols. Havana: Cultural.

Sahagún, Bernardino. [1583] 1982. *Introductions and Indices: Introductions, Sahagún's Prologues and Interpolations, General Bibliography, General Indices.* Edited by Arthur J. O. Anderson and Charles E. Dibble. Santa Fe, NM: School of American Research Press; Salt Lake City: University of Utah.

Sahlins, Marshall David. 1981. *Historical Metaphors and Mythical Realities: Structure in the Early History of the Sandwich Islands Kingdom.* Ann Arbor: University of Michigan Press.

Said, Edward W. 1994. *Culture and Imperialism.* New York: Knopf.

Saldaña, María Josefina. 2016. *Indian Given: Racial Geographies across Mexico and the United States.* Durham, NC: Duke University Press.

Sánchez Jiménez, Antonio. 2007. "Raza, identidad y rebelión en los confines del imperio hispánico: Los cimarrones de Santiago del Príncipe y *La Dragontea* (1598) de Lope de Vega." *Hispanic Review* 72, no. 2: 113–33.

———. 2008. "'Muy contrario a la verdad': Los documentos del Archivo General de Indias sobre *La Dragontea* y la polémica entre Lope y Antonio de Herrera." *Bulletin of Spanish Studies: Hispanic Studies and Researches on Spain, Portugal, and Latin America* 85, no. 5: 569–80.

Sanz Ayán, Carmen. 2004. *Estado, monarquía y finanzas: Estudios de historia financiera en tiempo de Los Austrias*. Madrid: Centro de Estudios Políticos y Constitucionales.

Schmitt, Carl. [1922] 1985. *Political Theology: Four Chapters on the Concept of Sovereignty*. Translated by George Schwab. Chicago: University of Chicago Press.

———. [1950] 2003. *The Nomos of the Earth in the International Law of the Jus Publicum Europaeum*. Translated by G. L. Ulmen. New York: Telos.

Schumpeter, Joseph Alois. 1942. *Capitalism, Socialism, and Democracy*. New York: Harper.

———. [1911] 1983. *The Theory of Economic Development: An Inquiry into Profits, Capital, Credit, Interest, and the Business Cycle*. New Brunswick, NJ: Transaction Books.

Schüssler, Rudolf. 2005. "On the Anatomy of Probabilism." In *Moral Philosophy on the Threshold of Modernity*, edited by Jilly Kraye and Risto Saarinen, 91–113. Dordrecht: Kluwer Academic.

Seed, Patricia. 1991. "Failing to Marvel: Atahualpa's Encounter with the Word." *Latin American Research Review* 26, no. 1: 7–32.

———. 1995. *Ceremonies of Possession in Europe's Conquest of the New World, 1492–1640*. Cambridge: Cambridge University Press.

Sepúlveda, Juan Ginés de. [1550] 1997. *Demócrates Segundo; Apología en favor del libro sobre las justas causas de la guerra*. Pozoblanco: Excmo. Ayuntamiento de Pozoblanco.

Shell, Marc. 1978. *The Economy of Literature*. Baltimore, MD: Johns Hopkins University Press.

———. 1982. *Money, Language, and Thought: Literary and Philosophical Economies from the Medieval to the Modern Era*. Berkeley: University of California Press.

———. 1993. *Children of the Earth: Literature, Politics, and Nationhood*. New York: Oxford University Press.

———. 2014. *Islandology: Geography, Rhetoric, Politics*. Stanford, CA: Stanford University Press.

Shklovsky, Victor. [1917] 1965. "Art as Technique." In *Russian Formalist Criticism: Four Essays*, translated by Lee T. Lemon and Marion J. Reis, 3–24. Lincoln: University of Nebraska Press.

Smith, Pamela H., and Paula Findlen. 2002. *Merchants and Marvels: Commerce, Science, and Art in Early Modern Europe*. New York: Routledge.

Spivak, Gayatri Chakravorty. 1996. *The Spivak Reader: Selected Works of Gayatri Chakravorty Spivak*. Edited by Donna Landry and Gerald M. MacLean. New York: Routledge.

Sprague, Jesse Rainsford. 1943. *The Romance of Credit*. New York, London: D. Appleton-Century Co.

Stern, Steve J. 1993. *Peru's Indian Peoples and the Challenge of Spanish Conquest: Huamanga to 1640*. Madison: University of Wisconsin Press.

Suárez, Francisco. [1597] 1965. *Disputationes Metaphysicae*. Hildesheim: Olms.

Sullivan, Ceri. 2002. *The Rhetoric of Credit: Merchants in Early Modern Writing*. London: Associated University Press.

Tambiah, Stanley Jeyaraja. 1990. *Magic, Science, Religion and the Scope of Rationality*. Cambridge: Cambridge University Press.

Tawney, R. H. 1922. *Religion and the Rise of Capitalism: A Historical Study*. Gloucester: Smith.

Teresa of Avila. [1588] 2007. *Libro llamado castillo interior, o las moradas*. Alexandria, VA: Alexander Street Press.

Thirst, Nigel. 1996. "The Making of Capitalist Time-Consciousness, 1300–1800." In *Human Geography: An Essential Anthology*, edited by John Agnew et al., 552–70. Cambridge: Blackwell.

Todorov, Tzvetan. 1984. *The Conquest of America: The Question of the Other*. Translated by Richard Howard. New York: Harper.

Tomlinson, Gary. 2007. *The Singing of the New World: Indigenous Voice in the Era of European Contact*. Cambridge: Cambridge University Press.

Townsend, Camilla. 2003. "Burying the White Gods: New Perspectives on the Conquest of Mexico." *American History Review* 108, no. 3: 659–87.

Trujillo Mena, Valentín. 1981. *La legislación eclesiástica en el Virreynato del Perú durante el siglo XVI: Con especial aplicación a la jerarquía y a la organización diocesana*. Lima: Editorial Lumen.

Tuan, Yi-fu. 1976. "Geopiety: A Theme in Man's Attachment to Nature and to Place." In *Geographies of the Mind: Essays in Historical Geosophy in Honor of John Kirtland Wright*, edited by David Lowenthal et al., 11–39. New York: Oxford University Press.

Valla, Lorenzo. [ca. 1440] 1998. *The Profession of the Religious and Selections from The Falsely-Believed and Forged Donation of Constantine*. Translated by Olga Zorzi Pugliese. Toronto: Centre for Reformation and Renaissance Studies.

Van Deusen, Nancy. 2015. *Global Indios: The Indigenous Struggle for Justice in Sixteenth-Century Spain*. Durham, NC: Duke University Press.

Vargas Llosa, Mario. 1987. "Latin America: Fiction and Reality." In *On Modern Latin American Fiction*, edited by John King, 1–17. New York: Farrar.

Varon Gabai, Rafael. 1980. *Curacas y encomenderos: Acomodamiento nativo en Huaraz siglos XVI–XVII*. Lima: Villanueva.

Vasari, Giorgio. [1550] 1998. *The Lives of the Artists*. Edited by Peter E. Bondanella. Translated by Julia Conaway. Oxford: Oxford University Press.

Vas Mingo, Milagros del. 1986. *Las capitulaciones de Indias en el siglo XVI*. Madrid: Cultura Hispánica.

Vega, Lope de. 1965. *Obras completas*. Edited by Joaquín de Entrambasaguas (vol. 1). Madrid: Consejo Superior de Investigaciones Científicas.

Verlinden, Charles. 1978. "Pagos y moneda en la América colonial." In *Dinero y crédito (siglos XVI y XIX): Actas del Primer Coloquio Internacional de Historia Económica*, ed. Alfonso Otazu, 325–34. Madrid: Artes Gráficas Benzal.

Vico, Giambattista. 2001. *New Science: Principles of the New Science Concerning the Common Nature of Nations*. London: Penguin.

Vilar, Pierre. 1978. "La noción de empresa y empresario desde los tiempos modernos a los contemporáneos." In *Dinero y crédito (siglos XVI y XIX): Actas del Primer Coloquio Internacional de Historia Económica*, edited by Alfonso Otazu, 241–48. Madrid: Artes Gráficas Benzal.

Vilches, Elvira. 2010. *New World Gold: Cultural Anxiety and Monetary Disorder in Early Modern Spain*. Chicago: University of Chicago Press.

Vitoria, Francisco de. [1532] 1963. *Las relecciones "De Indis" et "De iure belli."* Edited by Javier Malagón Barceló. Washington, DC: Unión Panamericana.

Voigt, Lisa. 2009. *Writing Captivity in the Early Modern Atlantic: Circulations of Knowledge and Authority in the Iberian and English Worlds*. Chapel Hill: University of North Carolina Press.

Wachtel, Nathan. 1973. *Sociedad e ideología: Ensayos de historia y antropología andinas*. Lima: Instituto de Estudios Peruanos.

Wagner, Henry Raup, and Helen Rand Parish. 1967. *The Life and Writings of Bartolomé de Las Casas*. Albuquerque: University of New Mexico Press.

Wallace, David. 2004. *Premodern Places: Calais to Surinam, Chaucer to Aphra Behn*. Malden, MA: Blackwell.

Wallerstein, Immanuel. 1974. *The Modern World-System I: Capitalist Agriculture and the Origins of the European World-Economy in the Sixteenth Century*. San Diego, CA: Academic Press.

Walton, Nicholas. 2015. *Genoa, "La superba": The Rise and Fall of a Merchant Pirate Superpower*. London: Hurst and Co.

Weber, Max. [1889] 2003. *The History of Commercial Partnerships in the Middle Ages*. Edited by Lutz Kaelber. Lanham, MD: Rowman & Littlefield.

———. 2009a. *From Max Weber: Essays in Sociology*. Edited by Hans Heinrich Gerth. Translated by C. W. Mills. New York: Routledge.

———. [1905] 2009b. *The Protestant Ethic and the Spirit of Capitalism: The Talcott Parsons Translation Interpretations*. Translated by Richard Swedberg. New York: Norton.

Wey-Gómez, Nicolás. 2008. *The Tropics of Empire: Why Columbus Sailed South to the Indies*. Cambridge, MA: MIT Press.

White, Hayden V. 1985. *Tropics of Discourse: Essays in Cultural Criticism*. Baltimore, MD: Johns Hopkins University Press.

———. 1987. *Metahistory: The Historical Imagination in Nineteenth-Century Europe*. Baltimore, MD: Johns Hopkins University Press.

———. 1999. *Figural Realism: Studies in the Mimesis Effect*. Baltimore, MD: Johns Hopkins University Press.

———. 2014. *The Practical Past*. Evanston, IL: Northwestern University Press.

Wimsatt, W. K., Jr., and M. C. Beardsley. 1946. "The Intentional Fallacy." *Sewanee Review* 54, no. 3: 468–88.

Wood, Neal. 1967. "Machiavelli's Concept of 'Virtù' Reconsidered." *Political Studies* 15, no. 2: 159–72.

Wright, John Kirtland. 1966. *Human Nature in Geography: Fourteen Papers, 1925–1965*. Cambridge, MA: Harvard University Press.

INDEX

Page numbers followed by *f* indicate a figure.

Aristotle, 5, 76, 132, 146–47, 198, 228n21, 231–32n10. *See also* metalepsis

asientos, 45–46, 57, 115; *de esclavos* (slave contracts), 227n10. *See also* contracts

ayllus, 187, 191, 210

Azores Islands, 77, 118, 126–28, 130

Badiou, Alain, 6, 224n5, 244n18

bankers, 32, 45–48, 122, 228n15, 228n19, 229n22, 236n12

beloved, the, 8, 12, 17, 58, 69–70, 76

biopower, 149, 205, 217, 221, 239n12

Bourdieu, Pierre, 12–13, 225nn10–11

branding, 93, 109, 113, 115, 142

Brevísima relación de la destruición de las Indias (Las Casas), 11, 141, 147, 199, 236n12; conscience in, 112; *conquista* and, 49, 238n9; excoriation of Spaniards in, 49, 60, 90, 99; legal justice and, 11, 151; narrative of, 120, 149; publication of, 122, 237n1; violence and, 141. See also *Apologética historia sumaria*; Las Casas, Bartolomé de

bureaucracy, 21–22, 48, 90, 141, 226n3, 234n25

Braudel, Fernand, 11, 24–25, 32, 57, 127, 229n23, 237n15

Byzantine novels, 114, 117–18

Canary Islands, 21, 119, 122, 125, 128–30, 138, 235n6

cannibalism, 9, 145, 153, 177, 238n6

capital, 10, 17, 46, 50, 120, 129, 134, 188, 205, 207, 217; campaigns, 122; dominion and, 39–40; flourishing of, 53–54; loans, 227n10; power and, 44, 140; processes of, 238n9; risk, 38; scalability and, 7; *societas* and, 55–56; sovereignty and, 43, 203; Spanish Empire and, 33, 36–39, 168, 171; usury and, 13; violence and, 5, 60. *See also* capital investment; *empresa*; venture capital

capital investment, 22, 24, 35, 37, 186; access to, 52; *cupiditas* and, 41; *encomenderos* and, 192–94; monarchic, 39, 46–47, 100, 118, 129, 229n22; returns to, 44, 101, 220. *See also* capital; general partners; interest; labor; limited partners; managerial expertise; *societas*; venture capital

capitalism: conquest and, 23, 38, 133–34, 207; evangelization and, 219; monarchic, 24, 39, 41, 56, 127 (*see also* Nunes Dias, Manuel); nature and, 123, 128, 132, 140; origins of, 11, 224n6, 229n23; production of, 226n4, 228n19; Weber on, 58. *See also* debt; interest; labor; merchants; venture capital

"Capitalocene," 11, 140

capitulaciones (contracts), 25, 54, 72, 95, 98, 171–72, 227n10, 228n16, 233n18

caritas, 12, 15, 73, 100, 220–21, 231–32n10, 232n14; *amor* and, 69–72, 99; conflation with *cupiditas*, 6, 154, 158, 173–74, 205, 209, 225n9; contradiction with *cupiditas*, 159, 219; forced reconciliation with *cupiditas*, 166, 232n15; Petrarchan, 71, 232n16. See also *Amor*; charity; Christian love; *cupiditas*

carried interest, 23, 35, 42, 46, 100, 103, 174, 200. *See also* capital investment; general partners; interest

NICOLE D. LEGNANI

is assistant professor of Spanish and Portuguese
at Princeton University. She is the translator of
Titu Cusi: A 16th-Century Account of the Conquest

CPSIA information can be obtained
at www.ICGtesting.com
Printed in the USA
LVHW081053020322
712398LV00005B/154